D0949883

Electrical Inspection Guidebook

Electrical Inspection Guidebook

John E. Traister

Reston Publishing Company
A Prentice-Hall Company
Reston, Virginia

Library of Congress Cataloging in Publication Data

Traister, John E

Electrical inspection guidebook.

Includes index.
1. Electric wiring—Inspection—United States—Handbooks, manuals, etc. I. Title.
TK152.T67 621.319'24 78-27006
ISBN 0-8359-1629-4

Reston Publishing Company
A Prentice-Hall Company
Reston, Virginia

10 9 8 7 6 5 4 3 2 1

Printed in the United States of America

Table of Contents

Preface

This book is designed to explain the "why's" as well as the "how-to's" involved in observation of electrical construction quality, giving the complete roles and responsibilities of the electrical inspector. Those involved in electrical inspection—architectural/engineering firms, government agencies, and municipal building departments, for example—will be able to find out exactly what is expected of them and how to carry out their duties properly.

Presenting handy checklists to work from, this book shows the reader exactly what should be reviewed and inspected in areas such as rough wiring, electric services, grounding, materials, wiring devices, underground wiring, overhead wiring, and electrical equipment. It further correlates construction inspection with successful management procedures, including use of documentation, change orders, progress payments, and construction scheduling with modern methods.

The chapters are arranged in a structural sequence intended to follow the way electrical systems are installed in buildings. Described are the inspection duties involved with each of these components. Explanations of OSHA (National Electrical Code) requirements as they apply to the inspector are also covered.

This book also fills the reader in on the inspector's role involving change orders, progress payments, documentation, and supervising testing of completed electrical systems.

John E. Traister

Chapter 1

Electrical Inspection

In nearly all but the very small electrical installations, contract specifications—as prepared by architects and engineers—require inspection or approval of the various items of material and the installation methods. Furthermore, all electrical construction is usually subjected to city, county, state, or power company inspection to determine if materials and equipment or installation methods conform to requirements of the National Electrical Code (NEC). Such inspections are made during the progress of roughing-in, at times determined by the extent to which the wiring system is to be concealed within the building structure and, finally, upon completion of the project.

It is not unusual to find two or more public authorities having inspection jurisdiction over a given job and occasionally, inspectors may not agree in their interpretation of the NEC requirements. When such a situation occurs, contact the National Fire Protection Association (NFPA); the National Electrical Code Committee has established interpretation procedures.

Most local inspection authorities have no jurisdiction over building construction on projects owned or leased by the federal government. Each of the various federal governmental agencies establishes its *own* inspection procedures and requirements covering both the type and quality of material and its installation, as well as its own design and inspection departments.

While all governmental agencies usually base their requirements on the NEC, they may differ slightly from the NEC's minimum requirements.

All electrical inspectors who anticipate working with such agencies should familiarize themselves with the inspection and installation regulations of the particular government agency having jurisdiction over any building or project they may be assigned.

1.1 QUALIFICATIONS OF AN ELECTRICAL INSPECTOR

A successful electrical inspector must have a thorough working knowledge of the National Electrical Code and have the ability to read and interpret blueprint (working drawings) and specifications. He should be able to compile reports and prepare correspondence. For these reasons, an electrical inspector should have the equivalent of at least a high school education, not necessarily a formal education, but the equivalent of 12 years of elementary and secondary schooling.

There is no substitute for experience, and all inspectors who are placed in responsible charge of electrical construction work should have had prior experience on construction work. The experience should demonstrate an intimate knowledge of electrical materials including trade names used in the field. For example, if an electrician refers to wiring methods as "Romex" and "BX", the inspector should know that these terms mean nonmetallic-sheathed cable and metal-clad cable respectively. He should also have a good knowledge of electrical practices, installation methods, techniques, equipment, and tools.

The ability to read and work regularly from working drawings and specifications and the ability to read and understand Federal specifications and commercial standards is absolutely necessary.

Probably the best way to acquire the experiences mentioned above is to actually work as an electrician on several projects. However, many successful electrical inspectors have gained this knowledge by working under experienced personnel.

Most electrical inspectors must be in good health and physically capable to meet the demands of the work. For example, he may be required to climb ladders or walk several flights of stairs to reach an area with work to be inspected. He may be required to climb down into manholes and then crawl through a tunnel containing electrical wiring that must be inspected.

If persons who anticipate working as an electrical inspector lack any of the above qualifications, they should take steps to rectify the situation, or possibly seek another profession.

1.2 ELECTRICAL INSPECTION DEFINED

Basically, electrical inspection is the observation of quality—quality of materials and quality of workmanship. The electrical inspector must have the ability to judge these qualities, detect any errors and omissions, and then make certain that discrepancies are corrected before any further work is carried out or before the job receives final approval.

1.3 THE ROLE OF THE ELECTRICAL INSPECTOR

The electrical inspector must be able to look upon and view critically the particular phase of the construction project to which he is assigned. He must closely follow the progression of each stage of the electrical construction and be alert to existing conditions as well as problems that might arise in the future. When the inspector notices through his periodic inspections that certain phases of the work are not being performed in accordance with the NEC or local ordinances, or when other problems occur, he must immediately report these errors, violations, or problems to the proper authorities for further action.

FIG. 1-1: The electrical inspector must be able to look upon and view critically all types of electrical construction projects to which he is assigned.

Those inspectors working for architectural/engineering firms must further make sure that all work is performed in accordance with the plans and specifications or any other architectural addenda.

In most cases, the inspector is not authorized (as an individual) to revoke, alter, substitute, enlarge, relax, or release any requirements of any specifications, plans, drawings, etc. In addition, the inspector is normally not authorized to approve or accept any segment of the work that is contrary to the construction documents. At no time should the individual inspector stop the construction work or interfere with the contractor's employees; this is a job for "management."

In effect, the inspector acts as liaison between the contractor and the owner or architect. How effective he can be in his role depends mainly on how he handles himself in this relationship. He must display knowledge, experience, integrity, ability, and the use of good judgement.

1.3.1 Specific Duties

The responsibilities of the electrical inspector employed by municipal building departments include the following:

1. Checking all electrical contractors working in the jurisdiction as to qualifications, valid licenses, permits, etc.

2. Checking the qualifications of those in responsible charge of a project if other than a qualified electrical contractor.

3. Reviewing contract drawings and specifications to insure that they meet with provisions set forth in the NEC, local ordinances, etc.

FIG. 1-2: One of the many jobs of an electrical inspector is to review contract documents to ensure that they meet with provisions set forth in the National Electrical Code.

4. Making periodic visits to the various projects to which the inspector is assigned to view critically the quality of materials, methods, and workmanship. These visits should be planned so as not to hold up the construction any more than is absolutely necessary.

5. Rejecting work that does not comply with NEC requirements or local ordinances.

6. Being alert to all construction work in a jurisdiction and making certain that applications have been filed by the contractor requesting periodic inspections.

Those inspectors working for architectural/engineering firms and governmental agencies will normally have the following responsibilities:

1. Making certain that the electrical contractor complies with the approved construction documents, that is, the working drawings, specifications, and contractual provisions for the project.

2. Monitoring the project to insure that the project progresses according to schedule, and reporting any problems.

3. Coordinating and monitoring material approvals and certain tests that may be required.

4. Interpreting construction documents and settling minor disputes.

5. Approving progress payments and checking all work against progress payment requests.

6. Approval of shop drawings, materials, samples, etc.

7. Monitoring change orders and making recommendations for approval or rejection.

From the above list, it is obvious that the electrical inspector has different responsibilities and authorities, dependent upon the organizational set-up under which he is working, and upon his own capabilities. Each inspector should understand his own level of technical knowledge and accept his responsibilities without overstepping his authority.

1.4 THE NATIONAL ELECTRICAL CODE (NEC)

Because of the potential fire and explosion hazards caused by the improper handling and installation of electrical wiring, certain rules in the selection of materials, quality of workmanship, and precautions for safety must be followed. In order to standardize and simplify these rules and provide some reliable guide for electrical construction, the National Electrical Code (NEC) was developed. The NEC, originally prepared in 1897, is frequently revised to meet changing conditions such as improved equipment and materials and new fire hazards. The Code is a result of the best efforts of electrical engineers, manufacturers of electrical equipment, insurance underwriters, firefighters, and other concerned experts throughout the country.

The NEC is now published by the NFPA (National Fire Protection Association), 470 Atlantic Ave., Boston, Mass. 02210. It contains specific rules and regulations intended to help in "the practical safeguarding of persons and property from hazards arising from the use of electricity. . . ." The NEC contains provisions considered necessary for safety. Compliance therewith and proper maintenance will result in an installation essentially free from hazard, but not necessarily efficient, convenient, or adequate for good service for future expansion of electrical use.

FIG. 1-3: The National Electrical Code Book—the electrical inspector's Bible.

The NEC has become the "Bible" of the electrical construction industry and anyone involved in electrical work—in any capacity—

must obtain an up-to-date copy of it, keep it handy at all times, and refer to it frequently.

1.5 LOCAL CODES AND ORDINANCES

A number of towns and cities have their own local electrical codes or ordinances. In general, these are based on, or are similar to, the NEC, but on certain classes of work they may have a few specific rules that are usually more rigid than the NEC.

In addition to the NEC and local ordinances of certain cities, there may exist with local power companies some special rules regarding location of service-entrance wires, meter connections, and similar details that must be satisfied before connection can be made to a building.

1.6 UNDERWRITERS' LABORATORIES

If approved wiring methods are used to install the electrical wiring in a building, but low quality materials are used, the complete installation may still be dangerous. Therefore, insurance companies have led the way to insure minimum standards of quality in electrical materials which, through experiment and experience, lead to a maximum of usefulness with the least amount of danger.

The American Insurance Association (AIA) has established a testing organization known as Underwriters' Laboratories (UL), Inc., with testing facilities in several locations throughout the United States. Manufacturers may submit samples of their products to these laboratories for testing before they are manufactured on a large scale. If the products pass the exhaustive tests in accordance with established standards, they are listed in the UL official published list and are then known as "Listed by Underwriters' Laboratories, Inc." Such UL approved items usually have a UL label attached directly to the product. In some cases, this UL "label" is molded or stamped into the merchandise—into bakelite, steel, or porcelain parts of wiring devices.

Underwriters' Laboratories is a nonprofit organization supported by the manufacturers who submit merchandise; that is, for testing the merchandise, Underwriters' Laboratories charges a fee, which pays for the inspectors' expenses and supports the laboratories in general.

A UL label assures the user that the manufacturer of the item has submitted to the laboratories for testing samples that were found to meet the required minimum safety standards. It is then presumed that the manufacturer will maintain the same quality in future production of the same item.

FIG. 1-4: Those products that pass the exhaustive tests of Underwriters' Laboratories Inc. usually have a UL label attached directly to the product.

An item approved by Underwriters' Laboratories, however, does not mean that the item is approved for all uses. Rather, a UL label means that the item or device as labelled is safe *only* for the purpose for which it was intended; NEC regulations must still be followed. Type NM cable, for example, cannot be used as a service-entrance cable just because it happens to bear the UL label.

The UL label furthermore means that the item meets the minimum safety requirements; one kind may meet the requirements, while another may far exceed them. Therefore, quality cannot really be judged from the UL label alone.

FIG. 1-5: This piece of NM cable is UL listed; however, the cable can be used for installations only as approved by the NEC.

Chapter 2
Construction Documents

When the construction of a building is contemplated, an architect (using consulting engineers) is usually commissioned to prepare the complete working drawings and specifications for the project covering the details of bidding and construction. The number of drawings will vary from only a few on smaller projects to many dozen full-size drawings on larger, more complex ones. In nearly all cases, regardless of the project size, the working drawings will fall into the following general groups.

Site Plans A site plan, provided on most building construction projects, shows the location of the building(s) on the property line and existing and new utilities such as electricity, water, sewer, and telephone. For some projects, contour lines are provided to show the lay of the land and other pertinent details.

Architectural Drawings The architectural drawings include floor plans showing the walls, partitions, doors, and windows for each floor level of the building; elevations of all exterior faces of the building; sectional views showing the various details of stairways, ceiling, and roof structures, and similar details. Large scale drawings, known as construction detail drawings, are also provided for special items like interior trim and cabinets.

Mechanical Drawings The mechanical drawings cover the complete design and layout of the heating, ventilating, air conditioning, plumbing, and related mechanical construction. They are normally prepared by consulting engineers working under the architect's supervision. These drawings include floor plans of the different levels showing all water pipes, sanitary pipes, air ducts, and other mechanical equipment, with cross-sections of duct-work as well as mechanical details.

Electrical Drawings The electrical drawings cover the complete design and layout of the wiring for lighting, power, signals and communications, special electrical systems, and related electrical equipment. These drawings sometimes include a site plan showing the location of the building on the property and the interconnecting electrical systems; floor plans showing the location of power outlets, lighting fixtures, panelboards, etc.; power-riser diagrams; a symbol list; schedules; schematic diagrams; and larger scale details where necessary.

2.1 BASIC ELECTRICAL DRAWINGS

In order to be able to "read" any of these drawings, one must first become familiar with the meaning of the various symbols, lines, and abbreviations used on the drawings and learn how to interpret the message conveyed by each one.

The types of electrical drawings are:

- Electrical construction drawings
- Single-line block diagrams
- Schematic wiring diagrams

Electrical construction drawings show the physical arrangement and views of specific electrical equipment. These drawings give all the plan views, elevation views, and other details necessary to construct the installation. Electrical construction drawings are used mainly by electrical equipment manufacturers, but the electrical inspector will sometimes run across electrical construction drawings like the one shown in Fig. 2-1. This type of construction drawing is normally used to supplement building electrical-system drawings for a special installation and is often referred to as an electrical detail drawing.

Electrical diagrams are drawings that are intended to show, in diagrammatic form, electrical components and their related connections. In diagrams, electrical symbols are used extensively to represent the various components. Lines are used to connect these symbols indicating the size, type, and number of wires necessary to complete the electrical circuit.

The electrical inspector will often come into contact with single-line block diagrams. These are used extensively by consulting engineering firms to indicate the arrangement of electrical ser-

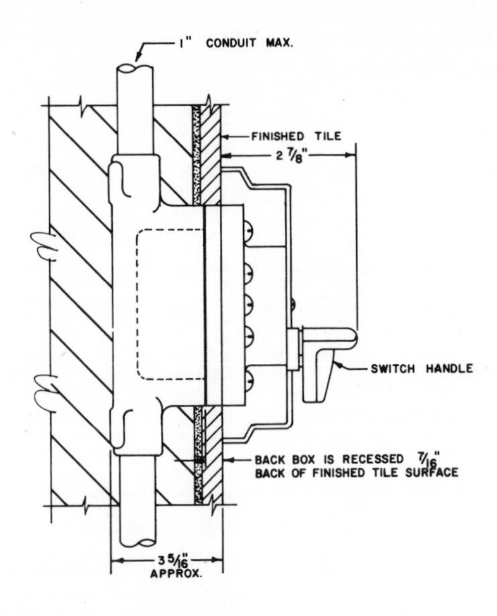

FIG. 2-1. A typical electrical detail drawing.

vices on electrical working drawings. The power-riser diagram in Fig. 2-2 for example is typical of such drawings. This particular drawing shows all of the panelboards and related equipment, as well as the connecting lines, to indicate the circuits and feeders. Notes are used to identify each piece of equipment, and to indicate the size of conduit necessary for each circuit or feeder, and the number, size, and type of insulation on the conductors in each conduit.

A schematic wiring diagram (Fig. 2-3) is similar to a single-line block diagram except that the schematic diagram gives more-detailed information and shows the actual size and number of wires used for the electrical connections.

Anyone involved in the electrical construction industry—in any capacity—frequently encounters all the above (three) types of electrical drawings. Therefore, it is very important for all involved in this industry to fully understand electrical drawings, wiring diagrams, and other supplementary information found in working drawings and in written specifications.

From the above paragraphs, it is evident that electrical drawings utilize symbols, lines, and notation to show the extent and requirements of a given electrical system. Most engineers and draftsmen use electrical symbols adopted by the United States of America Standards Institute (USASI). However, many of these sym-

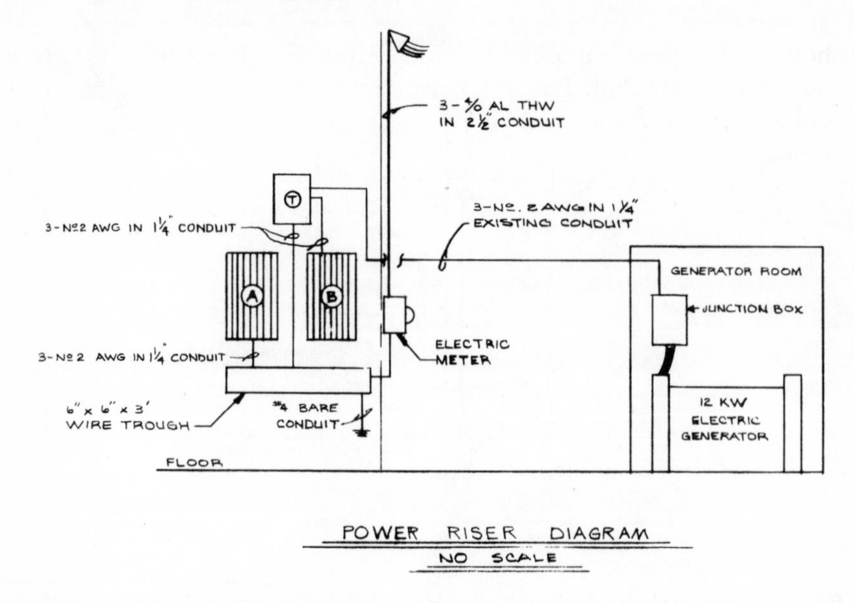

FIG. 2-2. Typical power riser diagram.

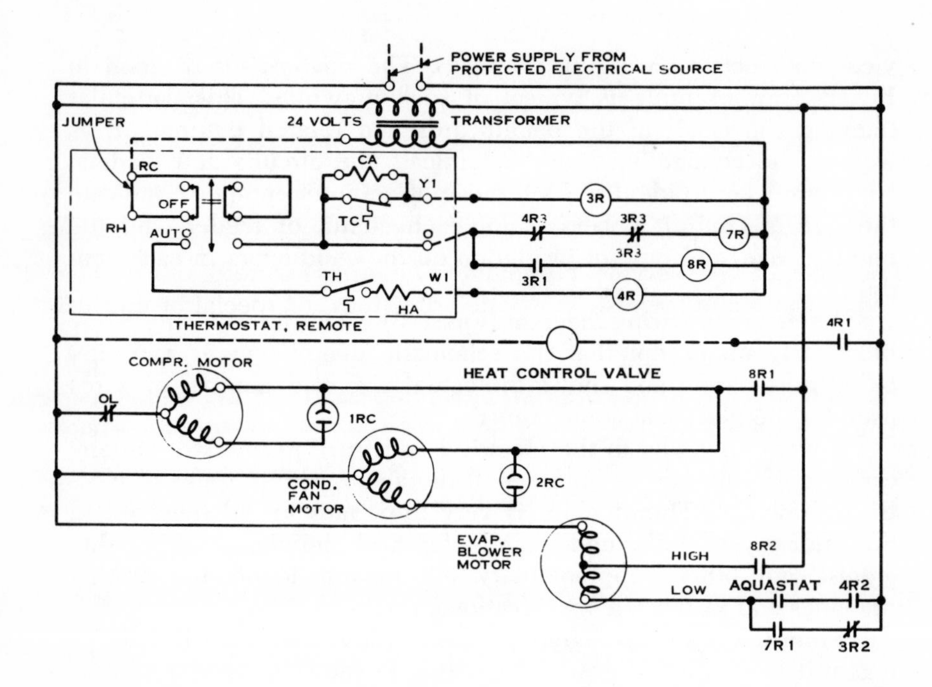

FIG. 2-3. Typical schematic wiring diagram.

bols are frequently modified to suit certain needs for which there is no standard symbol. For this reason, most drawings include a symbol list or legend as part of the drawings or the written specifications.

Figure 2–4 shows a list of electrical symbols prepared by the Consulting Engineers Council/U.S. and the Construction Specifications Institute, Inc. This list represents a good set of electrical symbols that are easily interpreted by workmen and sufficient for most applications.

2.2 PREPARATION OF ELECTRICAL DRAWINGS

The ideal electrical drawing should show in a clear, concise manner exactly what is required of the workmen. The amount of data shown on such a drawing should be sufficient, but not overdone. Unfortunately this is not always the case. The quality of electrical drawings vary from excellent, complete, and practical to just the opposite. In some cases, the design may be so incomplete that the electrical contractor or his estimator will have to supplement—if not completely design it—prior to estimating the cost of the project or beginning the installation of the electrical system.

In general, a good electrical drawing should contain floor plans of each floor of the building—one for lighting and one for power riser diagrams to diagrammatically show the service equipment, feeders, and communication equipment; schedules to indicate the components of the service equipment, lighting fixtures, and similar equipment; and large-scale detailed drawings for special or unusual portions of the installation. A legend or electrical symbol list should also be provided on the drawings in order to explain the meaning of every symbol and line used on the drawings. Anything that cannot be explained by symbols and lines should be clarified with neatly lettered notes or explained in the written specifications. The scale to which the drawings are prepared is also important; they should be as large as practical, and where dimensions are to be held to extreme accuracy, dimension lines should be added. Figure 2–5 shows a poorly prepared electrical drawing while Fig. 2–6 shows one of relatively good quality. In the former drawing, it is obvious that the electrical contractor will have to lay out or design portions of the system before it can be properly estimated, installed, or examined thoroughly by the electrical inspection department.

ELECTRICAL SYMBOLS

SWITCH OUTLETS

Single-Pole Switch — S

Double-Pole Switch — S_2

Three-Way Switch — S_3

Four-Way Switch — S_4

Key-Operated Switch — S_K

Switch and Fusestat Holder — S_FH

Switch and Pilot Lamp — S_p

Fan Switch — S_F

Switch for Low-Voltage Switching System — S_L

Master Switch for Low-Voltage Switching System — S_{LM}

Switch and Single Receptacle

Switch and Duplex Receptacle

Door Switch — S_D

Time Switch — S_T

Momentary Contact Switch — S_{MC}

Ceiling Pull Switch

"Hand-Off-Auto" Control Switch — HOA

Multi-Speed Control Switch — M

Push Button

RECEPTACLE OUTLETS

Where weather proof, explosion proof, or other specific types of devices are to be required, use the upper-case subscript letters. For example, weather proof single or duplex receptacles would have the uppercase WP subscript letters noted alongside of the symbol. All outlets should be grounded.

Single Receptacle Outlet

Duplex Receptacle Outlet

Triplex Receptacle Outlet

Quadruplex Receptacle Outlet

Duplex Receptacle Outlet-Split Wired

Triplex Receptacle Outlet-Split Wired

250 Volt Receptable Single Phase Use Subscript Letter to Indicate Function (DW-Dishwasher; RA-Range, CD - Clothes Dryer) or numeral (with explanation in symbol schedule)

250 Volt Receptacle Three Phase

Clock Receptacle — C

Fan Receptacle — F

Floor Single Receptacle Outlet

Floor Duplex Receptacle Outlet

Floor Special-Purpose Outlet *

Floor Telephone Outlet - Public

Floor Telephone Outlet - Private

Example of the use of several floor outlet symbols to identify a 2, 3, or more gang floor outlet:

Underfloor Duct and Junction Box for Triple, Double or Single Duct System as indicated by the number of parallel lines.

Example of use of various symbols to identify location of different types of outlets or connections for underfloor duct or cellular floor systems:

J

Cellular Floor Header Duct

*Use numeral keyed to explanation in drawing list of symbols to indicate usage.

FIG. 2-4. List of electrical symbols.

CIRCUITING

Wiring Exposed (not in conduit)	—E—
Wiring Concealed in Ceiling or Wall	
Wiring Concealed in Floor	
Wiring Existing*	
Wiring Turned Up	
Wiring Turned Down	
Branch Circuit Home Run to Panel Board.	2 1

Number of arrows indicates number of circuits. (A number at each arrow may be used to identify circuit number.)**

BUS DUCTS AND WIREWAYS

Trolley Duct***	T T
Busway (Service, Feeder, or (Plug-in)***	B B
Cable Trough Ladder or Channel***	C C
Wireway***	W W

PANELBOARDS, SWITCHBOARDS AND RELATED EQUIPMENT

Flush Mounted Panelboard and Cabinet***	
Surface Mounted Panelboard and Cabinet***	
Switchboard, Power Control Center, Unit Substations (Should be drawn to scale)***	
Flush Mounted Terminal Cabinet (In small scale drawings the TC may be indicated alongside the symbol)***	TC
Surface Mounted Terminal Cabinet (In small scale drawings the TC may be indicated alongside the symbol)***	TC

Pull Box (Identify in relation to Wiring System Section and Size)	
Motor or Other Power Controller (May be a starter or contactor)***	
Externally Operated Disconnection Switch***	
Combination Controller and Disconnection Means***	

POWER EQUIPMENT

Electric Motor (HP as indicated)	1/4
Power Transformer	
Pothead (Cable Termination)	
Circuit Element, e.g., Circuit Breaker	CB
Circuit Breaker	
Fusible Element	
Single-Throw Knife Switch	
Double-Throw Knife Switch	
Ground	
Battery	
Contactor	C
Photoelectric Cell	PE
Voltage Cycles, Phase	Ex: 480/60/3
Relay	R
Equipment (Connection (as noted)	

*Note: Use heavy-weight line to identify service and feeders. Indicate empty conduit by notation CO (conduit only).
**Note: Any circuit without further identification indicates two-wire circuit. For a greater number of wires, indicate with cross lines, e.g.:

3 wires; 4 wires, etc.

Neutral wire may be shown longer. Unless indicated otherwise, the wire size of the circuit is the minimum size required by the specification. Identify different functions, of wiring system. e.g., signalling system by notation or other means.
***Identify by Notation or Schedule

FIG. 2-4 continued

REMOTE CONTROL STATIONS FOR MOTORS OR OTHER EQUIPMENT

Pushbutton Station	PB
Float Switch - Mechanical	F
Limit Switch - Mechanical	L
Pneumatic Switch - Mechanical	P
Electric Eye - Beam Source	
Electric Eye - Relay	
Temperature Control Relay Connection (3 Denotes Quantity.)	R_3
Solenoid Control Valve Connection	S
Pressure Switch Connection	P
Aquastat Connection	A
Vacuum Switch Connection	V
Gas Solenoid Valve Connection	G
Flow Switch Connection	F
Timer Connection	T
Limit Switch Connection	L

LIGHTING

	Ceiling	Wall
Surface or Pendant Incandescent Fixture (PC = pull chain)	Type, Watts	Switch, PC, circuit
Surface or Pendant Exit Light		
Blanked Outlet	B	B
Junction Box	J	J
Recessed Incandescent Fixtures		
Surface or Pendant Individual Fluorescent Fixture		
Surface or Pendant Continuous-Row Fluorescent Fixture (Letter indicating controlling switch)	A; 1 ← Fixture No.; 100 ← Wattage; Symbol not needed at each fixture	
*Bare-Lamp Fluorescent Strip		

ELECTRIC DISTRIBUTION OR LIGHTING SYSTEM, AERIAL

- Pole**
- Street or Parking Lot Light and Bracket**
- Transformer**
- Primary Circuit**
- Secondary Circuit**
- Down Guy
- Head Guy
- Sidewalk Guy
- Service Weather Head**

ELECTRIC DISTRIBUTION OR LIGHTING SYSTEM, UNDERGROUND

Manhole**	M
Handhole**	H
Transformer Manhole or Vault**	TM
Transformer Pad**	TP
Underground Direct Burial Cable (Indicate type, size and number of conductors by notation or schedule)	
Underground Duct Line (Indicate type, size, and number of ducts by cross-section identification of each run by notation or schedule. Indicate type, size, and number of conductors by notation or schedule.	
Street Light Standard Fed From Underground Circuit**	

*In the case of continuous-row bare-lamp fluorescent strip above an area-wide diffusing means, show each fixture run, using the standard symbol; indicate area of diffusing means and type by light shading and/or drawing notation.
**Identify by Notation or Schedule

FIG. 2-4 continued

SIGNALLING SYSTEM OUTLETS

INSTITUTIONAL, COMMERCIAL, AND INDUSTRIAL OCCUPANCIES

I Nurse Call System Devices (any type)

Basic Symbol.

(Examples of Individual Item Identifiction Not a part of Standard)

Nurses' Annunciator (add a number after it as ⊢① 24 to indicate number of lamps) ⊢①

Call Station, single cord, pilot light ⊢②

Call Station, double cord, microphone speaker ⊢③

Corridor Dome Light, 1 lamp ⊢④

Transformer ⊢⑤

Any other item on same system - use numbers as required. ⊢⑥

II Paging System Devices (any type)

Basic Symbol.

(Examples of Individual Item Identification. Not a part of Standard)

Keyboard ⊢◇1

Flush Annunciator ⊢◇2

2-Face Annunciator ⊢◇3

Any other item on same system - use numbers as required ⊢◇4

III Fire Alarm System Devices (any type) including Smoke and Sprinkler Alarm Devices

Basic Symbol.

(Examples of Individual Item Identification. Not a part of Standard)

Control Panel ⊢[1]

Station ⊢[2]

10" Gong ⊢[3]

Pre-signal Chime ⊢[4]

Any other item on same system - use numbers as required. ⊢[5]

IV Staff Register System Devices (any type)

Basic Symbol

(Examples of Individual Item Identification. Not a part of Standard)

Phone Operators' Register

Entrance Register - flush

Staff Room Register

Transformer

Any other item on same system - use numbers as required.

V Electric Clock System Devices (any type)

Basic Symbol.

(Examples of Individual Item Identification. Not a part of Standard)

Master Clock

12" Secondary - flush

12" Double Dial - wall mounted

18" Skeleton Dial

Any other item on same system - use numbers as required.

VI Public Telephone System Devices

Basic Symbol.

(Examples of Individual Item Identification. Not a part of Standard)

Switchboard 1

Desk Phone 2

Any other item on same system - use numbers required. 3

FIG. 2-4 continued

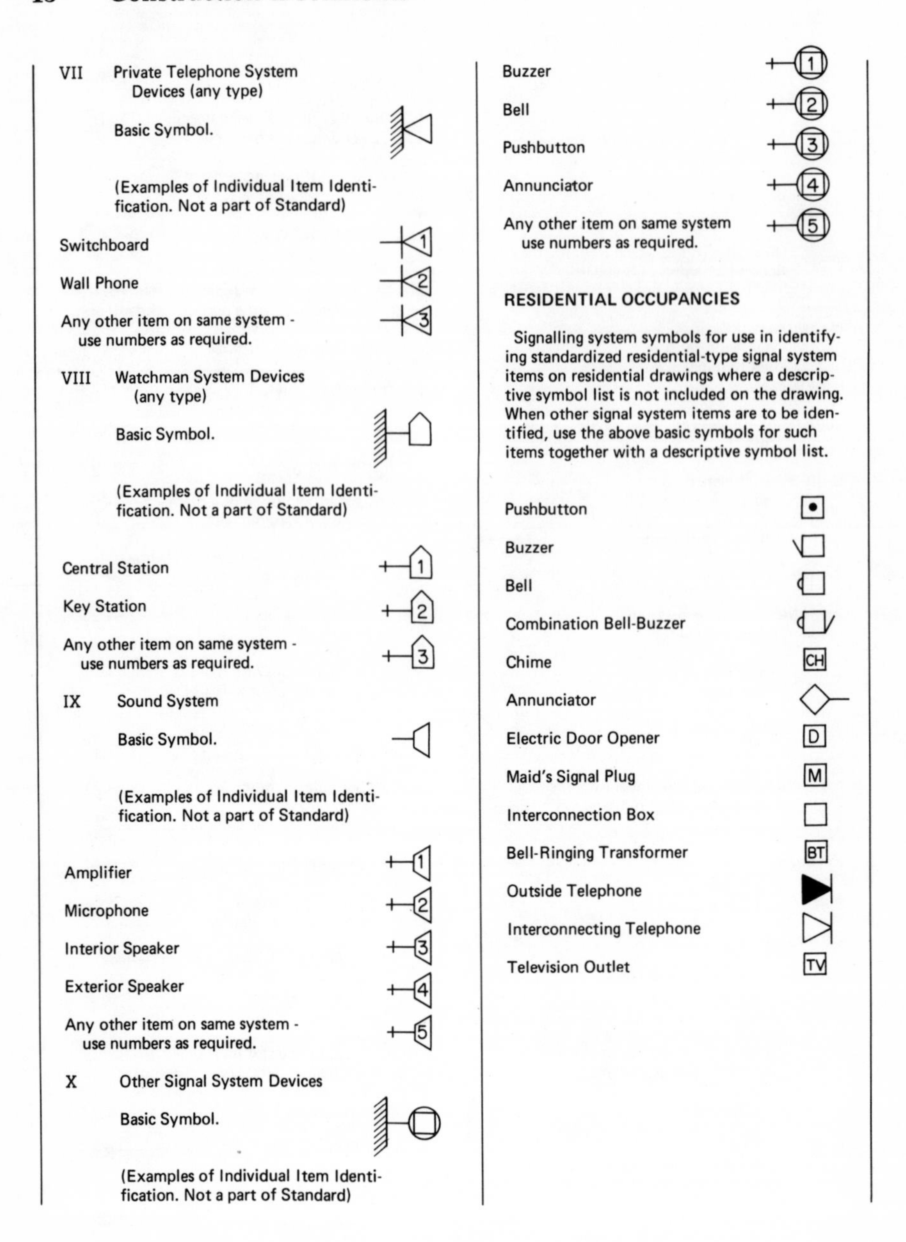

FIG. 2-4 continued

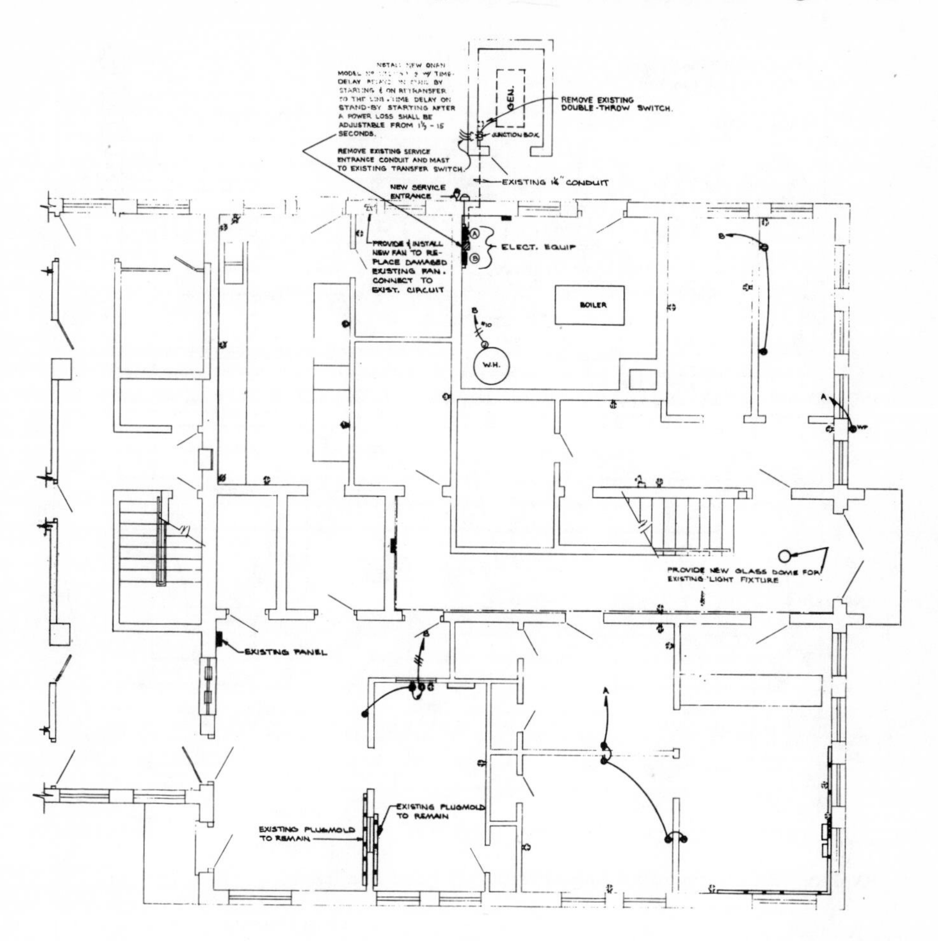

FIG. 2-5. A poorly prepared electrical drawing with many necessary details left out.

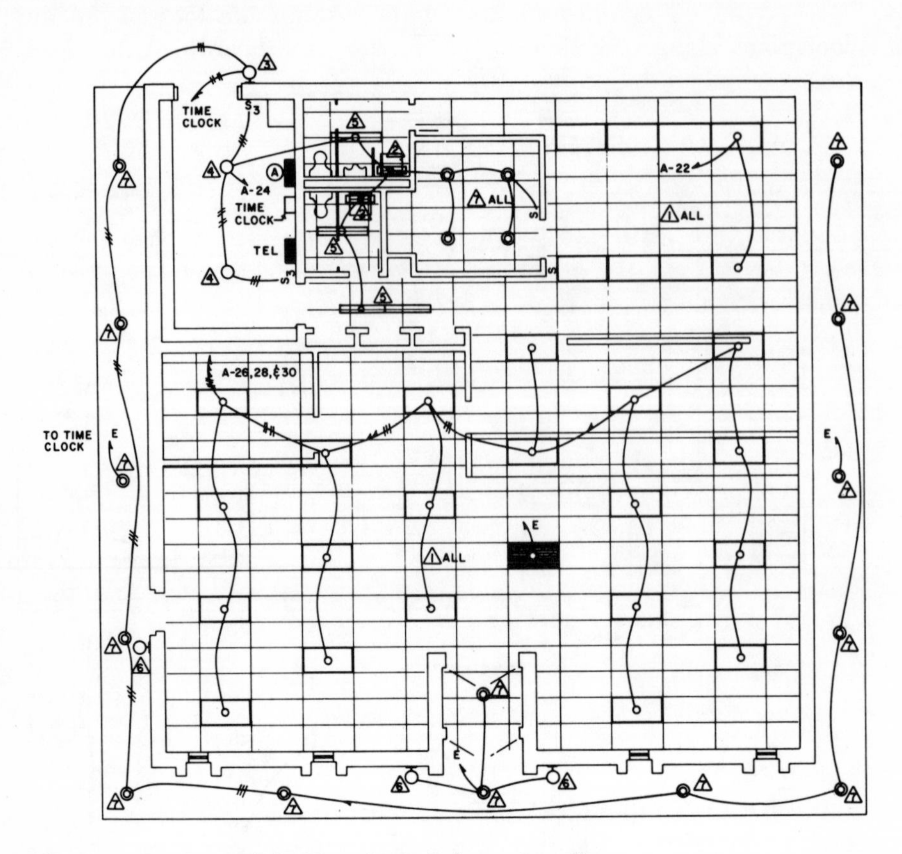

FIG. 2-6. An electrical drawing of relatively good quality.

The following steps are usually necessary in preparing a good set of electrical working drawings and specifications:

1. The engineer or electrical designer meets with the architect and owner to discuss the electrical needs of the building in question and to discuss various recommendations made by all parties.

2. Once the data in No. 1 above are agreed upon, an outline of the architect's floor plan is drawn on tracing paper and then several prints of this floor plan outline are made.

3. The designer or engineer then calculates the required power and lighting requirements for the building and sketches them on the prints.

floor plans along with lighting and power panelboards. Again these are sketched on the prints.

5. Circuit calculations are made to determine wire size and overcurrent protection and then reflected on the drawings.

6. After all the electrical loads in the entire building have been determined, the main electric service and related components (transformers, etc.) are selected and sketched on the prints.

7. Next schedules are made to identify various pieces of electrical equipment.

8. Wiring diagrams are made to show the workmen how various electrical components are to be connected. An electrical symbol list is also included to identify the symbols used on the drawings.

9. Various large-scale electrical details are included, if necessary, to show exactly what is required of the workmen.

10. Written specifications are then made to give a description of the materials and the installation methods.

2.3 DRAWING SCHEDULES

A schedule, as related to electrical drawings, is a systematic method of presenting in tabular form notes or lists of equipment on a drawing. When properly organized and thoroughly understood, schedules save much valuable time not only for the draftsmen, but also for the electrical contractor and his personnel in preparing the estimate and installing the equipment in the field.

For example, the lighting-fixture schedule in Fig. 2–7 lists the fixture type from letters or numbers on the drawings. The manufacturer and catalog number of each fixture is included along with the number, size, and type of lamp for each. The "Volts" and "Mounting" columns follow, and the column on the extreme right is for special remarks, such as giving the mounting height for a wall-mounted fixture.

Sometimes schedules are omitted on the drawings and the information is contained in the written specifications instead. Combing through page after page of written specifications can be time consuming and many workmen do not always have access to the

FIXT. TYPE	MANUFACTURER'S DESCRIPTION	LAMPS		VOLTS	MOUNTING	REMARKS
		NO.	TYPE			

FIG. 2-7. Typical lighting fixture schedule.

specifications while working, whereas they usually do have access to the working drawings at all times.

The schedules in Figs. 2-8 through 2-10 are typical of those used on electrical drawings by consulting engineers.

2.4 SECTIONAL VIEWS

Sometimes the construction of a building is difficult to show with the regular projection views normally used on electrical drawings. If too many broken lines are needed, for example, to show hidden objects in buildings or equipment, the drawings become confusing and difficult to read. Therefore, in most cases, building sections are shown on working drawings to clarify the construction. To better understand a building section, imagine that the building has been cut into sections as if with a saw. The floor plan of a building in Fig. 2-11 shows a sectional cut at point A-A. This cut is then shown in Fig. 2-12.

In dealing with sections, one must be able to visualize considerably. Some sections are very easy to read while others are ex-

ELECTRIC-HEAT SCHEDULE					
HT'R TYPE	MANUFACTURER'S DESCRIPTION	DIMENSIONS	VOLTS	MOUNTING	REMARKS

FIG. 2-8. Electric heat schedule.

PANELBOARD SCHEDULE

PANEL Nº.	TYPE CABINET	PANEL MAINS			BRANCHES					ITEM-FED OR REMARKS
		AMPS	VOLTS	PHASE	1P	2P	3P	PROT	FRAME	
A	SURFACE	100 A	120/240 V	1φ, 3W	7	—	—	20 A	70 A	NEW CIRCUITS (RECEPTS & LIGHTS)
SQ. "D" TYPE NQO W/ 100A MAIN BKR.					13	—	—	20 A	70 A	SPARES
B	SURFACE	100 A	120/240 V	1φ, 3W	3	—	—	20 A	70 A	NEW RECEPT - CIRCUIT
SQ. "D" TYPE NQO W/ 100A MAIN BKR.					—	3	—	60 A	70 A	EXIST. PNL.S "B"-"C" & "D"
					—	1	—	60 A	70 A	ELECT. RANGE CIRCUIT
					—	1	—	30 A	70 A	WATER HEATER
					1	—	—	20 A	70 A	BOILER
					—	1	—	20 A	70 A	SPARE OR OTHER EQUIP.
					8	—	—	20 A	70 A	SPARES

NOTE:
REPLACE ALL EXISTING NON-GROUNDING TYPE RECEPTACLES W/ P&S CAT. NO. 5242, 3-WIRE GROUNDING TYPE RECEPTACLES, OR BRYANT AS INDICATED IN SPECS.
ALL DASHED RECEPTS ARE EXISTING.
ALL SOLID-LINE RECEPTACLES ARE NEW.
PULL NEW FEEDER IN EXISTING CONDUIT TO 3-EXISTING SUB-PANELS.

FIG. 2-9. Panelboard schedule.

PROJECT: ____________ DATE: ____________

JOB: ____________ BY: ____________

INTERCOMMUNICATION SYSTEM SCHEDULE

ROOM NAME	ITEM	I.D. No.	1	2	3	4	5	6	7	8	9	10	11	12	13	14	15	16	17	18	19	20	21	22	23	24	25	26	REMARKS
		1																											
		2																											
		3																											
		4																											
		5																											
		6																											
		7																											
		15																											
		16																											
		17																											
		18																											
		19																											
		20																											
		21																											
		22																											
		23																											
		24																											
		25																											
		26																											

FIG. 2-10. Intercommunication schedule.

tremely difficult and there are no set rules for determining what a section will look like. For example, a piece of rigid conduit, cut vertically, will have the shape of a rectangle; cut horizontally, it will have a round or circular appearance; cut on the slant, it will be an ellipse.

A cutting plane line (Fig. 2-13) has arrowheads to show the direction in which the section is viewed. Letters such as A-A, B-B, etc., are normally used with the cutting-plane lines to identify the cutting plane and the corresponding sectional views.

2.5 WIRING DIAGRAMS

Complete schematic wiring diagrams are used infrequently on the average set of electrical working drawings—only in unique and highly complicated systems like control circuits—but it is important to have a thorough understanding of them when the need to interpret arises.

Components in schematic wiring diagrams are represented by symbols, and every wire is either shown by itself or included in an

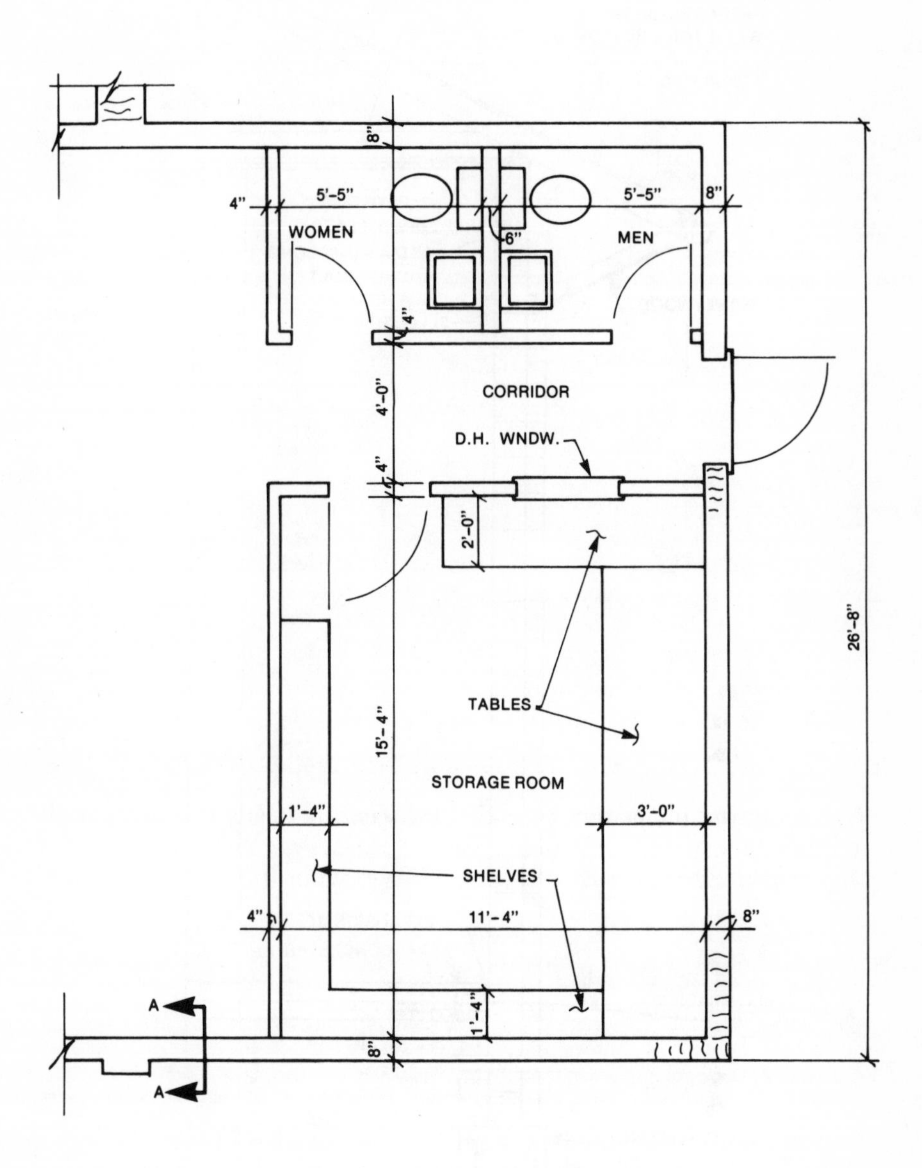

FIG. 2-11. Floor plan of a building showing a sectional cut at point A-A.

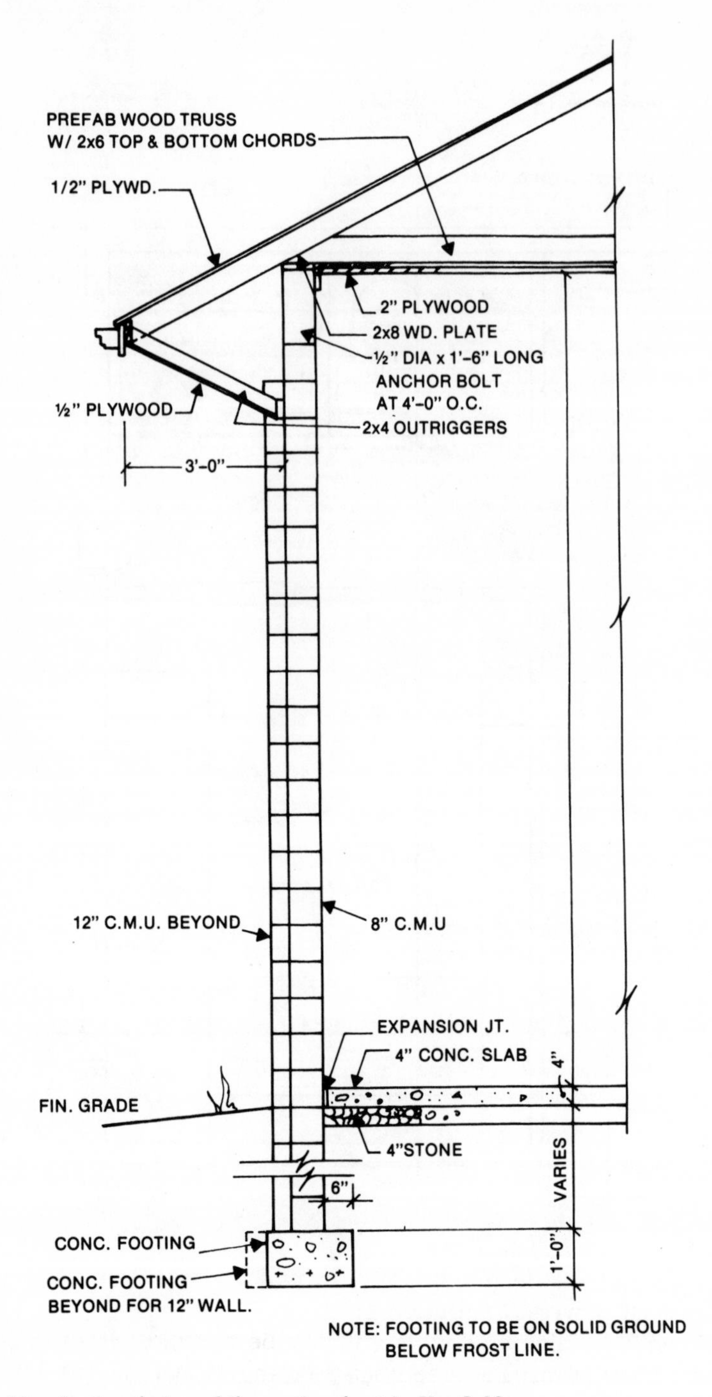

FIG. 2-12. Sectional view of the sectional cut in Fig. 2-11.

FIG. 2-13. Typical cutting plane line.

assembly or several wires that appear as one line on the drawing. Each wire in the assembly, however, is numbered when it enters and keeps the same number when it emerges to be connected to some electrical component in the system. Figure 2–14 shows a complete schematic wiring diagram for a three-phase, AC magnetic motor starter. Note that this diagram shows the various devices (in symbol form) and indicates the actual connections of all wires between the devices.

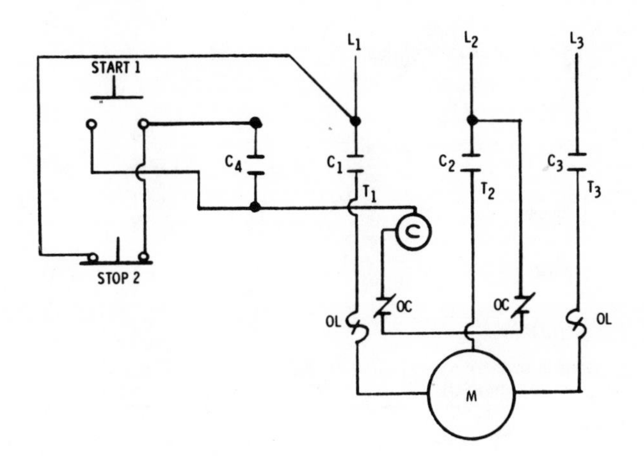

FIG. 2-14. Schematic wiring diagram for a 3-phase, ac magnetic motor starter.

Figure 2–15 gives a list of electrical wiring symbols commonly used for single-line schematic diagrams; single-line diagrams are simplified versions of complete schematic diagrams. Figure 2–16 shows the use of these symbols in a typical single-line diagram of an industrial power-distribution system.

Power-riser diagrams will probably be the most frequently encountered diagrams on electrical working drawings for building construction. Such diagrams give a picture of what components are

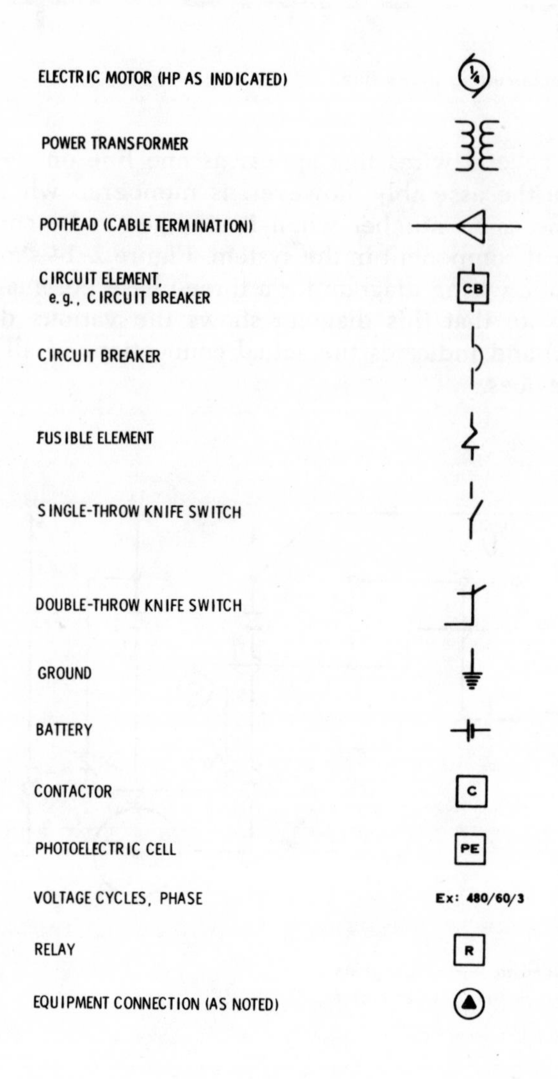

FIG. 2-15. List of electrical wiring symbols commonly used for single-line schematic diagram.

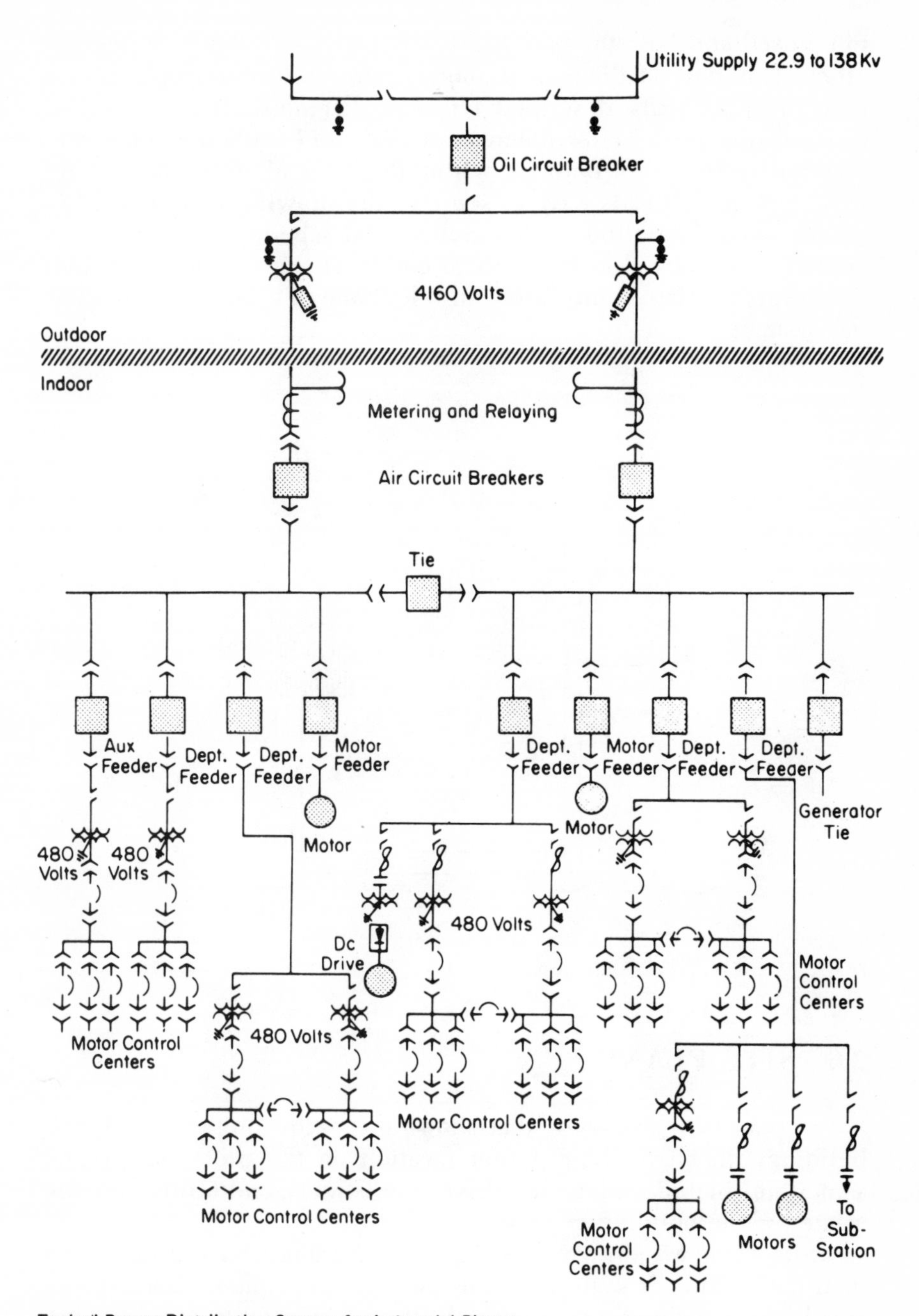

Typical Power Distribution System for Industrial Plants

FIG. 2-16. Use of the symbols in Fig. 2-15 in a typical single-line diagram of a power distribution system.

to be used and how they are to be connected in relation to one another. This type of diagram is more easily understood at a glance than the previously described types of diagrams. As an example, compare the power-riser diagram in Fig. 2-17 with the schematic diagram in Fig. 2-18. Both are wiring diagrams of an identical electrical system, but it is easy to see that the drawing in Fig. 2-17 is greatly simplified although a supplemental schedule is required to give all necessary data required to construct the system. Such diagrams are also frequently used on telephone, TV, alarms, and similar systems.

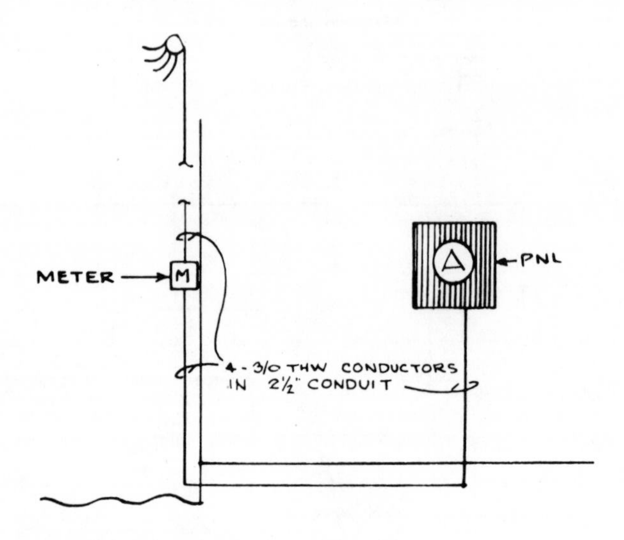

FIG. 2-17. A power riser diagram.

2.6 SITE PLANS

A site plan is a plan view that shows the entire property with the buildings drawn in their proper location on the plot. Such plans sometimes include sidewalks, driveways, streets, and utility systems related to the building or project.

Site plans are drawn to scale using the engineer's scale rather than the architect's scale used for most building plans. Usually, on small lots, a scale of, say, 1 in. = 10 ft or 1 in. = 20 ft is used. This means that 1 in. (actual measurement of the drawing) is equal to 10 ft, 20 ft, and so forth, on the land itself.

In general building-construction practice, it is usually the

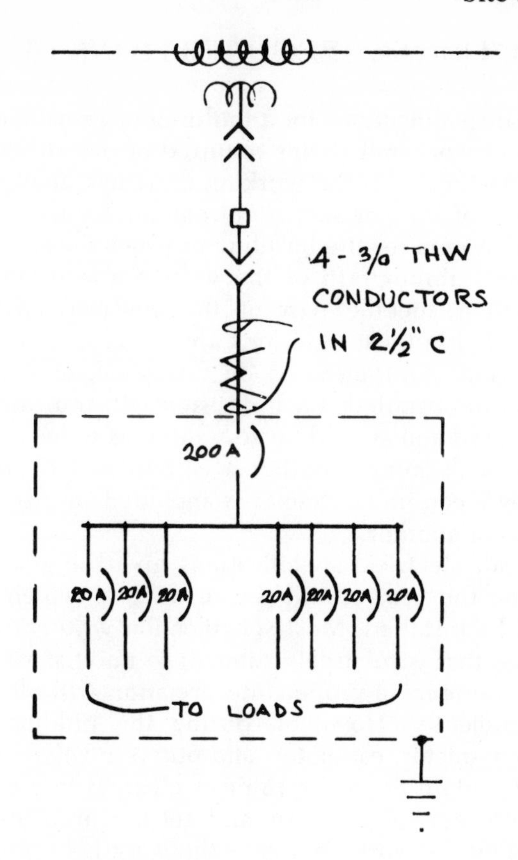

FIG. 2-18. A schematic diagram of the power riser diagram shown in Fig. 2-17.

owner's responsibility to furnish the architect with property and topographic surveys, which are made by a certified land surveyor or civil engineer. These surveys will show:

- all property lines
- existing utilities and their location on or near the property
- the direction of the land slope
- the condition of the land (woody, swampy, etc.)

The architects and engineer then use this site plan to incorporate all new utilities. The electrical contractor will then be concerned with the electrical distribution lines, the telephone lines, and possible cable TV lines, especially if they are to be installed underground.

2.7 ELECTRICAL SPECIFICATIONS

The electrical specifications for a building or project are the written descriptions of work and duties required of the architect, engineer, or owner. Together with the working drawings, these specifications form the basis of the contract requirements for the construction of the electrical system for the building or project.

Divisions 1 through 16 of the written specifications cover requirements of a specific part of the construction work on the project. Included in these divisions are the type and grade of materials to be used, equipment to be furnished, and the manner in which it is to be installed. Each division will indicate the extent of the work covered and should be so written as to leave absolutely no doubt in anyone's mind whether a certain part of the work to be performed by a certain contractor is included in one section of the specifications or another.

In general, electrical specifications give the grade of materials to be used on the project and the manner in which the electrical system shall be installed. Most specification writers use an abbreviated language that is relatively difficult to understand at first. However, with experience a proper interpretation will often be possible with little difficulty. However, during the bidding stage of the project the contractor, estimator, and others involved in the project should make certain that everything is clear. If it is not, contact the architectural or engineering firm and get the problem cleared up—prior to bidding the work—not after the work is in progress.

The following is an outline of the various sections normally included in Division 16 of the written construction specifications.

16010 General Provisions The General Provisions of the electrical specifications consist of a selected group of considerations and regulations that apply to all sections of this Division. Other items included in this section have been discussed previously.

16100 Basic Materials and Methods The contractor should look for clauses in this section that establish a means of identifying the type and quality of materials and equipment to be used in the project's electrical system. This section should further establish the accepted methods of installing the various materials.

16200 Power Generation Items covered in this section will vary from job to job, but usually cover equipment used for emergency or standby power facilities—the type used to take over essential electrical service during a normal power failure.

16300 Power Transmission This section deals mainly with high voltage (over 600 volts) power transmission circuits and will be included in projects constructed on government reservations and large industrial sites. For other projects, this type of work is usually handled by the local power company.

16400 Service and Distribution Power distribution facilities (under 600 volts) are covered in this section of the electrical specifications by descriptive paragraphs or clauses covering selected related equipment items.

16500 Lighting This section of the specifications normally covers general conditions relating specifically to the selected lighting equipment to insure that all lighting equipment is furnished and installed exactly as the designer selected and specified. Further clauses will establish the quality and type of all lighting fixtures, accessories, lamps, etc. Methods of installation will also be included in most sets of electrical specifications.

16600 Special Systems Such items as lightning protection equipment, special emergency light and power systems, storage batteries, battery charging equipment, and perhaps cathodic protection are examples of items that could be covered in this portion of the specifications.

16700 Communications Equipment items that are interconnected to permit audio or visual contact between two or more stations, or to monitor activity and operations at remote points are basically the items found in this portion of the specifications. Most clauses will be designed around a particular manufacturer's equipment stating what items will be furnished and what is expected of the system once it is installed and in operation.

16850 Heating and Cooling Due to working agreements between labor unions, most heating and cooling equipment, with few exceptions, are installed by workmen other than electricians and the requirements are usually covered in Division 15 of the specifications. However, in certain cases, the electrical contractor will be responsible for installing certain pieces of the equipment, especially on residential and apartment projects. This section of the electrical specifications will cover all necessary details pertaining to the equipment and the installation when such a condition exists.

16900 Controls and Instrumentation As the name implies, this section of the specifications covers all types of controls and instrumentation used on a given project. Some of the items covered will include recording and indicating devices, motor control centers, lighting control equipment, electrical interlocking devices and applications, control of electric heating and cooling, limit switches, and numerous other such devices and systems.

2.8 COMMON CONFLICTS

Electrical inspectors should always be on the alert for conflicts between working drawings and the written specifications, or between Division 16 and other Divisions of the construction documents. Such conflicts occur particularly when:

1. Architects or engineers use standard or prototype specifications and attempt to apply them to specific working drawings without any modification.

2. Previously prepared standard drawings are to be changed or amended by reference in the specifications only; the drawings themselves are not changed.

3. Items are duplicated in both the drawings and specifications and then an item is amended in one or the other and overlooked on the other contract document.

An example of item No. 3 would be where a power-riser diagram is shown on a sheet of the working drawings, showing diagrammatic locations of all panelboards and related service equipment to be used on the project. The written specifications list all of the panelboards including their contents (fuses, circuit breakers, etc.) in a panelboard schedule. If another panel must be added for the project at a later date, prior to the job going out for bids, it will most often be added to the power-riser diagram on the drawings, but such a change is often overlooked in the written specifications, especially if it is a last minute change. For this reason, it is best not to duplicate items in the specifications and in the drawings. Rather, determine the proper place for the information to be listed, and then indicate it in one or the other of the construction documents—not in both.

In such instances, it is the responsibility of the person in charge of the project to ascertain which takes precedent over the other—that is, the drawings or the specifications. When such a condition exists, the matter must be cleared up, preferably before the work is installed, in order to avoid added cost to either the owner, architect/engineer, or the contractor.

The electrical inspector must become familiar with the drawings and specifications, as these construction documents and the National Electrical Code will form the basis upon which the inspection is made. The inspector must be prepared for the fact that errors and omissions will be discovered, and must be duly corrected. When such errors are detected in the drawings or specifications, they should immediately be called to the attention of those in charge, either directly or through the proper channels so that appropriate action may be taken at the earliest possible time. When errors are pointed out, diplomacy should be used, to cause the least embarrassment.

The electrical inspector is on the job to be objective and to enforce the contract, NEC regulations, and other regulations pertaining to good workmanship and approved materials. In the case of differences between the building codes, plans, specifications, federal laws, state laws, local ordinances, industry standards, utility company regulations, fire insurance carrier's requirements, and the contract documents, the most stringent usually will govern. The inspector should promptly notify in writing the architect/engineer or others in charge of any such difference.

Chapter 3
Job Mobilization

This chapter has been designed with the primary objective of outlining practical and efficient preliminary job mobilization techniques that will serve as a guide to the electrical inspector.

3.1 PRELIMINARY CONFERENCE

Once a contract has been awarded, either by negotiation or competitive bid, the architect/engineer usually calls a meeting with all successful contractors on the project, including the electrical contractor. At this meeting, the drawings and specifications (including all addenda) are discussed and scheduled and manpower requirements will be agreed upon. This initial period of organization is also one of organization for the inspection team. The meeting will give the inspectors an idea of how they will need to schedule their time to be on the job so as not to hold up job progress.

Subjects of interest will include delivery and storage of materials, temporary power requirements, procedures for claims for extra work, and so forth. Those electrical inspectors responsible for verifying material delivered to the job site (to receive progress payments) should take notes of the dates the materials are due. This will enable the inspector to be on the job site at the dates given to check the material against the contract specifications. In most cases, however, the electrical materials will be late in arriving, especially for such items as panelboards and some lighting fixtures. Therefore, it is a good idea to check with the contractor (or have him check with the inspection department) to see if the materials are arriving on schedule before visiting the job site.

The preliminary conference is a good time for the electrical inspector to inform the electrical contractor what the electrical inspector expects in the way of workmanship, performance, methods of submitting shop drawings, and applying for permits. It is also a good time to check the contractor's liability insurance.

FIG. 3-1. The preliminary conference is a good time for the electrical inspector to inform the parties involved as to what he expects in the way of workmanship, performance, and the like.

3.2 BONDING

Normally the architect or owner of the building will require that the electrical contractor (and all other contractors) submit a performance bond to insure that the contractor will complete the work according to the plans and specifications. Should the contractor default, the company issuing the bond will be responsible for completing the project, usually by hiring another contractor.

Inspectors working for architectural/engineering firms must protect the owner from paying too much in progress payments because if this is allowed to happen, the bonding company can refuse to provide all of the funds necessary to finish the project. Therefore, when a progress payment request is submitted by the electrical contractor, it should state the percentage of work completed, what materials are in place, and what materials are stored on the job site. The inspector should then verify all such claims by an actual job site visit before approving the payment of the request. If care is taken in this phase of the construction, the owner will have little trouble obtaining all of the funds necessary to complete the project should the electrical contractor default.

Most performance bonds also guarantee the owner that the electrical work will be done on time and that all bills incurred will be paid. Due to the risks involved, surety companies are very careful in issuing surety bonds and many factors are considered prior to underwriting a performance and payment bond. Some considerations include:

1. Pass performance of the electrical contractor.

2. Financial status of the electrical contractor, including credit references and secondary assets.

3. Experience of the electrical contractor and his key personnel.

4. Size of organization and the number of tools and equipment owned by the electrical contractor.

All necessary information should be forwarded to the bonding companies well in advance of the anticipated bond requirements, as no agent is in a position to underwrite bonds without complete information at hand as explained above.

Sometimes an electrical contractor will be late in obtaining the required bonding and it then becomes the inspector's responsibility to take action, either extending the period of time a certain number of days or by canceling the contract. Most problems arise in the case of newer contractors who still lack some knowledge of certain adminstrative techniques, especially if the project is a great deal larger than ones performed previously. It is recommended that the electrical inspector work closely with such contractors to help them get on the right track, as everyone has to start some place.

3.3 PERMITS

Once the contract has been awarded, the electrical inspector should make certain that the successful contractor has obtained all of the required building permits. Some contractors wait until the last minute to obtain a permit and then find that the project is held up because certain information needed from the contractor is missing and the inspection department cannot issue the permits without it. Such information could be properly prepared working drawings, shop drawings, or a progress schedule, depending upon the authority having jurisdiction over the project.

Once the proper plans and specifications are received along with an application for the required building permits, the inspection department should promptly review the documents and if everything is in order, issue the permits without delay.

If a new contractor is performing a project in the inspection department's jurisdiction, the inspection department should show him courtesy in notifying the contractor of all requirements—that is, insurance required, bonds, personal qualifications, and so forth. This should be done as soon as possible; then, it the contractor lacks any of the requirements, he will have sufficient time to qualify without delaying the project.

Application for Zoning and Building

Date ________ Type Permit: New Const. ____ Permit No. ________
Addition ____
Demolition ____
Alteration ____
Mobil Home ____

Property Owner: Name ________ Phone No. ________
Address ________

Contractor: Name ________ Reg. No. ________
Electrical ________ Reg. No. ________
Plumbing ________ Reg. No. ________

Zoning: Street ________ Zoned ________ Map No. ________
(Refer to Deed) Lot No. ________ Block No. ________ Subdivision ________
Set Back ________ Right ________ Left ________ Rear ________
Previous Owner ________ Date Acquired ________

Building: Single Family ________ Duplex ________ Townhouse ________ Units ________ Other ________
Building Size ________ Ft. x ________ Ft. Finished Rooms ________
Exterior Walls: Brick ________ Siding ________ Other ________
Interior Walls: Drywall ________ Plaster ________ Paneling ________
Floors: Hardwood ________ Carpet ________ Concrete ________ Other ________
Roof Type & Material ________
No. Stories ________ Type Heat ________ Type Sewage Disposal ________
Source of Water ________ No. Bath Rooms ________
Type Air Conditioning ________ No. Fireplaces ________
Valuation ________ Fee ________

I hereby agree to comply with all provisions of "Uniform State Wide Building Code" (BOCA), the "Zoning Ordinance" and other ordinances or policies as adopted by the Town of

Signature of Applicant ________

Zoning Administrator ________
Building Inspector ________
Remarks: ________

A sketch showing location of structure on approved plat must accompany this application.

The Building Inspector shall be notified before footings or slabs are poured and before framing enclosure, and final inspection by calling the Town of

FIG. 3-2. A building permit is often required before the construction may begin.

3.4 INSURANCE

The insurance needs of electrical contractors vary somewhat from those of general contractors. The insurance needs are normally taken care of by the insurance agent, but the electrical inspector should know something about insurance coverage in order to approach his work more intelligently. Those types of insurance that concern the electrical inspector will normally fall into two distinct categories:

1. Liability for injury to other persons or for damage to their property.

2. Loss, damage to, or destruction of the contractor's property or personnel.

Most of the electrical inspector's concern will be with the contractor's liability insurance. In fact, most inspecting authorities require that certificates of liability insurance be filed with them before permits are issued or work begins. Bodily injury to individuals or damage to their property can result in legal claims against an electrical contractor in astronomical amounts. If the contractor does not have sufficient coverage, both the electrical contractor and the person whose property was damaged can loose.

3.4.1 General Liability Insurance

General liability insurance can be carried either on a scheduled liability form or on a Comprehensive General Liability Form; both pay on behalf of the electrical contractor. In general, the policy will pay all sums that the contractor may become legally obligated to pay as damages resulting in bodily injury to any member of the public, or as damages to his property caused by accident arising from the contractor's operations at all locations, including the contractor's office, warehouses, and so forth. In addition to paying a judgment awarded up to the policy limits, the insurance company agrees to provide the following benefits in most cases:

1. Defend any suit alleging injury.

2. Pay all premiums on bonds to release attachments or premiums on appeal bonds up to the policy limit.

3. Pay all costs of investigation, all costs assessed, and interest accruing on any judgment.

4. Pay for immediate medical and surgical fees necessary following an accident.

5. Reimburse the contractor for his expenses in litigation except loss of earnings.

Some of the items most policies do not include are:

1. Automobiles.

2. Liability to an employee while in the course of his employment or obligation payable under any workmen's compensation act.

3. Personal injuries other than bodily injuries that include such things as libel, slander, disparagement, defamation, invasion of privacy, and false arrest.

4. Injury to or destruction of property owned, used, rented to, or in the care or control of the insured.

3.5 INSPECTION DEPARTMENT MOBILIZATION

The electrical inspection department must familiarize itself with every project in its jurisdiction concerning the electrical installation and its relation to general and other specialty construction, both as to construction interrelationships and installation techniques. The following is a suggested guide to electrical inspectors in becoming familiar with a particular project.

1. Completely review all drawings and specifications as far in advance of the starting of construction as possible. This review should include specifications relating to the general building construction as well as all other specialty craft work. Particular attention should obviously be paid to the scope of work required under the electrical contract.

2. Completely review the progress schedules of both the general and electrical contractor.

3. Review any electrical installation progress and manpower schedules that may have been prepared for the job.

4. Whenever possible, the electrical inspector responsible for the given project should attend the preliminary conferences prior to beginning the work.

3.5.1 Review of Drawings and Specifications

When the drawings and specifications for a project are received by the inspection team, they should be reviewed from the following standpoints:

1. Extent of the over-all building, project, or facility to be provided.

2. Types of buildings or other construction involved.

3. Types and extent of mechanical and other special facilities involved.

4. Extent of the over-all electrical system to be provided.

5. General wiring methods required.

6. Types of electrical equipment, panelboards, lighting fixtures, etc.

7. Special related work.

The ability to visualize working drawings comes largely with experience. New inspectors with little actual construction experience can usually learn and make effective use of the basic over-all job electrical inspection functions. However, actual on-the-job experience should be obtained whenever possible.

The work of inexperienced workers should be thoroughly checked until they have completely mastered it, not only to be certain that the electrical installations have been installed properly, but in order to point out constructively to the inspector any errors he may have missed or improvements that can be made in the future.

Chapter 4
Branch Circuit Rough-In

In general, branch circuits include all outlet boxes, 1-in. and smaller conduit with fittings, and types of cable and building wire up to and including No. 8 AWG that is used to feed lighting, receptacles, small power, and communication systems.

Branch circuits may be installed *exposed, concealed, in earth, in concrete pours, in wood frame members, secured to steel structures,* and in numerous other localities. The type of building construction will depend to a great extent on the time the electrical inspector must visit the job site. For example, when a raceway system is run exposed on a masonry wall, the electrical inspector may delay the job site inspection by several days without holding up the progress of the job. On the other hand, Type NM cable run through wood studs in a residence, or rigid metal conduit buried in a concrete pour, for example, must be scheduled more thoroughly to avoid holding up the work of the installing of dry wall over the wiring or pouring the concrete.

The items covered in this chapter are designed to acquaint the electrical inspector with the knowledge of correct inspection techniques for inspecting branch circuit rough-in wiring of all types.

4.1 RESIDENTIAL ROUGH-IN

In all but the very largest homes (over 4000 sq ft), the electrical inspector should be able to complete the rough-in inspection of the electrical wiring in one visit to the project site, unless of course errors are found. Then an additional visit is required to make certain that the corrections have been made. Another visit normally referred to as the "final inspection" should complete the inspection of the average residential building, except when errors are found.

The electrical contractor should have all wiring that will later be concealed completed prior to the inspector's visit. If not, the job must be held up until such work has been completed. At times, on very large homes, the inspecting authority will allow a certain portion of the home to be covered before all of the concealed wiring is installed in other sections of the home. This is to allow the work of the other trades to continue should the rough-in wiring take longer than usual to install.

A completely roughed-in residential building should include the following:

1. All outlet boxes properly secured to the building structure—that is, outlet boxes for receptacles, wall switches, lighting fixtures, junction boxes, and the like.

2. All concealed wiring feeding the outlet boxes should be in place and properly secured, and all splices made in the outlet boxes. Wiring that will be partially exposed and partially concealed should also be run completely for the rough-in inspection.

3. Flush-mounted panelboards, electric fans, electric heaters, and other flush-mounted equipment should be mounted and all wires connected to the housing of the equipment. The wires do not necessarily need to be connected to the equipment terminals as long as they are accessible for connection later on. For example, in the case of a flush-mounted panelboard, usually only the housing will be installed and all cables and conduits connected to this housing. The loose wires are then left inside the empty housing until the wall covering is applied and finished. Once finished, the panel interiors (circuit breakers, fuse blocks, cable terminals, etc.) are installed, and the wiring is connected to their respective terminals. Unless the panel is energized prior to the final inspection, the panel cover is usually left off until the final inspection.

FIG. 4-1. Such items as flush-mounted electric heaters should be mounted and all wires connected to the housing prior to the rough-in inspection.

4. Service conductors from flush-mounted panels should be installed as well as service conductors running through a concealed area. If a conduit raceway system is used, only the raceway or conduit needs to be installed during the rough-in inspection; the conductors may be pulled in later.

5. If the exterior finish of the house is completed at the time the rough-in inspection is requested, many electrical inspectors prefer to have the meter base mounted, service entrance conductors installed, and the electrical system properly grounded. However, in most cases, such as with brick veneered homes, the rough wiring will be ready for inspection long before the exterior of the house is completely finished.

6. In most cases, heavy appliances such as electric ranges, electric clothes dryers, and electric water heaters will not be installed until sometime after the house is completed. If the feeders for these items are to be installed in areas that will later be concealed, the wiring should be installed with the load side terminating in a junction box or an approved power outlet.

4.1.1 Residential Wiring Methods

In nearly all instances, the major portion of wiring for residential branch circuits will consist of either type NM cable (nonmetallic sheathed) or type AC cable (armored cable). Type UF (underground feeder) cable is used extensively for underground wiring for outside lights and for feeding buildings that are not attached to the home itself. Rigid conduit and EMT (electrical metallic tubing) are sometimes used for service entrance masts and for a few other limited applications in the home, but such wiring methods are insignificant compared to most of the wiring.

Since type NM cable is the least expensive, this wiring method will be the most often found by the electrical inspector in residential wiring systems. This type of cable may be used in all types of

FIG. 4-2. The major portion of residential wiring will consist of type NM cable.

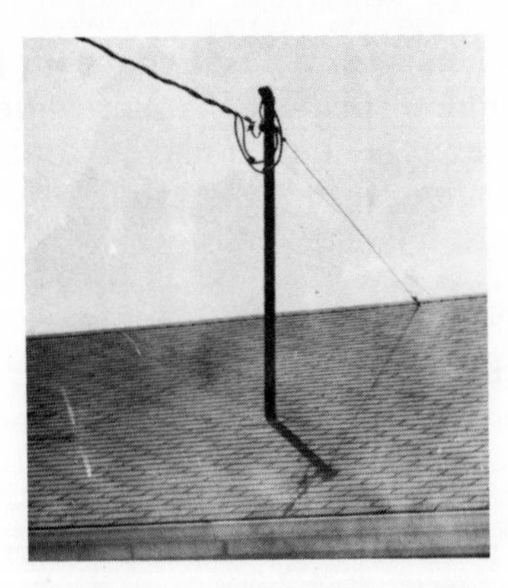

FIG. 4-3. Where a service mast is required, rigid metallic conduit is normally used.

dwelling units not exceeding three floors above grade. However, it shall not be installed where it will be exposed to corrosive fumes or vapors (found in a shop, for example), or embedded in masonry, concrete, or plaster.

When inspecting an NM cable installation, the inspector should make certain that all cable is secured in place at intervals not exceeding $4\frac{1}{2}$ ft and within 12 in. from every cabinet, box, or fitting. The staples, straps, or other fasteners must be designed and installed so as not to injure the cable. Where the cables pass through a floor, they must be protected from damage by conduit, guard strips, or other approved means; this protection should extend at least 6 in. above the floor.

FIG. 4-4. When inspecting NM cable installations, one of the requirements is that all cable be secured within 12 in. of every outlet box; the cable passing through the drilled hole in the stud above the box (as shown here) is considered to be "secured."

Where the cable is run at right angles with joists in unfinished basements, cables containing two No. 6 or three No. 8 conductors may be secured directly to the lower edges of the joists. Smaller cables, however, must be run either through bored holes in joists or on running boards. Of course, where cables are run parallel to the joists, cables of any size must be secured to the sides or faces of the joists.

Type AC cable is permitted for use in dry locations only. This type of cable is not approved for direct burial in the earth. The required supports for this cable are the same as those for type NM cable. The electrical inspector should make certain that all bends in the cable have a radius of the curve of the inner edge not less than 5 times the diameter of the cable. Furthermore, an anti-short bushing must be installed at every termination point of this cable. The NEC further requires that such bushings will be visible for inspection when the cable is connected to outlet boxes or cabinets.

Where run across the top of floor joists, or within 7 ft of floor or floor joists across the face of rafters or studding, or in accessible attics and roof spaces, the cable shall be protected by substantial guard strips that are at least as high as the cable. Where this space is not accessible by permanent stairs or ladders, protection shall be required only within 6 ft of the nearest edge of the scuttle hole or attic entrance. Where cable is carried along the sides of rafters, studs, or floor joists, neither guard strips nor running boards shall be required.

The electrical inspector should observe the wiring method used to feed major appliances. Because of the relatively heavy loads of these appliances, many residential contractors use type SE (service entrance) cable for feeders. If the grounded conductor of the SE cable is insulated, it may be used in the same way that type NM cable is used. However, if the grounded conductor is not insulated, the cable shall not be used within a building except for the following ways:

1. As a branch circuit to supply only a range, wall-mounted oven, counter-mounted cooking unit, or clothes dryer as covered in Section 250-60 of the NEC.

2. As a feeder to supply only those other buildings that are on the same premises.

3. Where the fully insulated conductors are used for circuit wiring and the uninsulated conductor is used for equipment grounding purposes—not as a grounded or neutral conductor.

Type UF cable may be used underground, including direct burial in the earth, as a feeder or branch circuit cable where provided with overcurrent protection of the rated ampacity as required in Section 339–4 of the NEC. Type UF cable is also permitted for interior wiring in wet, dry, or corrosive locations under the recognized wiring methods of the NEC. This type of cable should not be used as service entrance cables or embedded in poured cement, concrete, or aggregate, or where exposed to direct rays of the sun unless the covering is approved for the purpose.

In some areas of the home, especially for modernization work in existing homes, surface metal raceway is used. When used, the raceway, elbows, fittings, and outlet boxes must be of the same manufacture and design for use together. The size of surface metal raceway must be approved for the number and size of wires needed. Connections shall be made to other types of raceways in an approved manner with fittings manufactured for the purpose and application (see Figure 4-5). Where combination metal raceways are installed for signal, lighting, and power circuits, each system shall be run in separate compartments clearly identified and maintaining the same relative position throughout the system.

Electrical inspectors working for architectural/engineering firms will be required not only to inspect the methods in which the cable is installed, but also to make certain that the materials comply with the written specifications. A typical specification for wires and cables appears below:

16120 Wires and Cables

(A) Wire and cable shall meet all standards and specifications applicable, and shall be in conformance with the latest edition of the NEC. Insulated wire and cable shall have size, type of insulation, voltage, and manufacturer's name permanently marked on outer covering at regular intervals not exceeding four feet. Wire and cable shall be delivered in complete coils or reels with identifying tags, stating size, type of insulation, etc.

(B) Wire and cable shall be suitably protected from weather and damage during storage and handling, and shall be in first class condition after installation.

(C) Conductors shall be soft drawn copper, ASTM B3 for solid wire, ASTM B8 for stranded conductors. Conductor wire sizes shall be American Wire Gauge (AWG)—No. 6 and larger of stranded construction, No. 8 and smaller of solid construction.

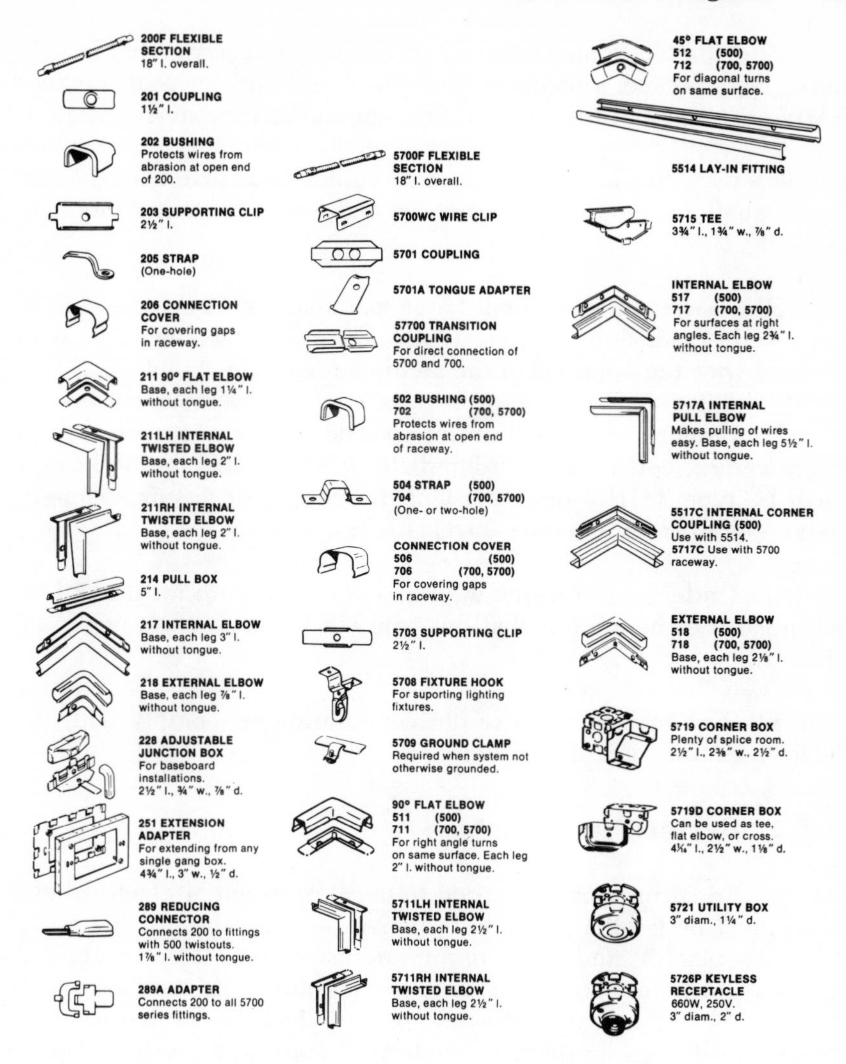

FIG. 4-5. Surface metal raceway and some related fittings.

(D) Wire and cable shall be factory color coded with a separate color for each phase and neutral used consistently throughout the system. Color codings shall be as required by the NEC.

(E) All conductors shall be rated 600 volts, unless otherwise specified or shown on the drawings, or for electronic or communication use.

(F) Conductors for lighting, receptacle, and power branch circuits, feeders, and subfeeders size No. 1 and smaller shall be type THW heat and moisture-resistant thermoplastic insulated.

(G) Conductor for feeders and subfeeders size No. 1/0 and larger shall be type RHW moisture and heat resistant rubber insulated.

(H) Wire and cable shall be as manufactured by ________ or approved equal. Substitution of wire and cable manufacturer shall be only with the approval of the architect/engineer.

(I) Branch circuits within all electric heater elements such as electric duct coils, baseboard radiation, and cabinet unit heaters shall be type THHN heat resistant, thermoplastic insulated, maximum operating temperature 90 °C (194 °F).

(J) Underground feeder and branch circuit wire for direct burial in earth or in conduit shall be type UF for use in wet or dry locations.

(K) For any specific use not covered above, comply with the NEC in conductor use.

4.1.2 Residential Outlets

The electrical inspector required to check residential electrical systems must be familiar with outlet box capacities, the spacing of duplex receptacles, and other requirements of the NEC. In general, the maximum numbers of conductors permitted in standard outlet boxes are listed in Table 370–6 (a) of the NEC. These figures apply where no fittings or devices, such as fixture studs, cable clamps, switches, or receptacles are contained in the box and where no grounding conductors are part of the wiring within the box. Obviously, in all modern residential wiring systems there will be one or more of these items contained in the outlet box. Therefore, where one or more of the above mentioned items are present, the number of conductors shall be one less than shown in the tables; an additional deduction of one conductor must be made for each strap containing one or more devices; and a further deduction of one conductor shall be made for one or more grounded conductors entering the box. A 3-in. × 2-in. × 2¾-in. box for example, is listed in the table as containing a maximum number of six No. 12 wires. If the box contains cable clamps in a duplex receptacle, two wires will have to be deducted from the total of six—providing for only four No. 12 wires.

Article 210–23(b) of the NEC specifically states the minimum requirement for the location of receptacles in residential buildings:

> *"In every kitchen, family room, dining room, breakfast room, living room, parlor, library, den, sun room, bedroom, recreation room, or similar rooms, receptacle outlets shall be installed so that no point along the floor line in any wall space is more than 6 ft, measured horizontally, from an outlet in that space, including any wall space 2 ft or more in width and the wall space occupied by sliding panels in exterior walls. The wall space afforded by fixed room dividers, such as free-standing bar-type counters, shall be included in the 6-ft measurement."*

In kitchen and dining areas a receptacle outlet must be installed at each counter space wider than 12 in. Countertop spaces separated by range tops, refrigerators, or sinks shall be considered as separate counter top spaces. Receptacles rendered inaccessible by appliances fastened in place or appliances occupying dedicated space shall not be considered as these required outlets.

Receptacle outlets, shall, insofar as practicable, be spaced equal distances apart. Receptacle outlets in floors shall not be counted as part of the required number of receptacle outlets unless located close to the wall.

At least one wall receptacle outlet shall be installed in the bathroom adjacent to the basin location. This receptacle shall be a 120-volt (15- or 20-ampere) receptacle and have a ground-fault circuit interrupter protection for personnel.

Check for the installation of at least one duplex receptacle outside the building. Also, at least one receptacle outlet in addition to any provided for laundry equipment shall be installed in each basement and in each attached garage.

At least one wall switch controlled lighting outlet must be installed in every habitable room, and in bathrooms, hallways, stairways, and attached garages as well as outdoor garages. One lighting outlet must also be installed in the following spaces:

- attic
- crawl space
- utility room
- basement

4.2 COMMERCIAL BUILDINGS

Branch circuit wiring in commercial buildings can be identical to those found in residential buildings. However, in most cases the main difference will be in the wiring method; that is, metal-clad cable will normally replace Type NM cable, and EMT or rigid conduit will replace Type SE or similar cable used for feeders in residential construction.

Branch circuit wiring in commercial buildings, extending from the panelboards or load centers to the various outlets, is not confined to any particular part of the building structure and is placed in the concrete floors, walls, partitions, and ceilings, exposed and concealed. The outlet boxes are secured to the building structure when the wiring is exposed and placed in the building when the wiring is concealed. The electrical inspector should make certain that the means of supporting these boxes comply with the NEC and that they are the proper type and size for the use.

In concealed wiring, the depth at which the faces of the boxes are installed with respect to the rough building structure depends upon the final finish and depth of outlet box covers used. The NEC requires that outlet boxes for enclosing flush devices be completely enclosed on back and sides. In walls or ceilings of concrete, tile, or other noncombustible material, boxes and fitting must be installed so that the front edge of the box or fitting will not set back of the finished surface more than 1/4 in. In other materials, the boxes and fittings must be flush with the finished surface or project.

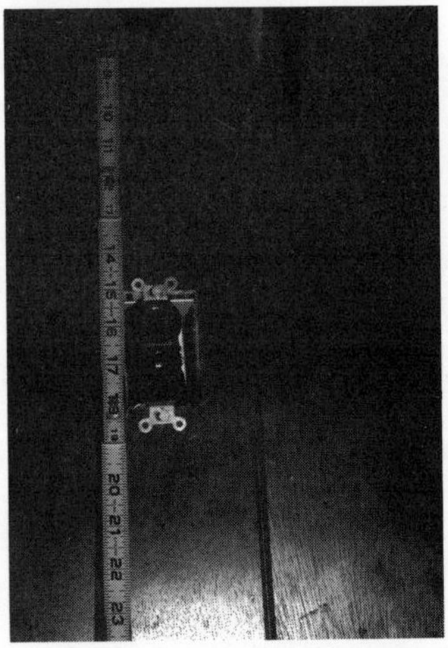

FIG. 4-6. When outlet boxes are installed in walls with combustible finishes, the outlet box must be flush with the finished surface as shown here.

Rigid conduit or EMT raceways are installed or concealed within the building structure in a manner depending upon the option of the user (provided all NEC requirements are met), or the relation of the size of the raceway to the dimensions of the portion of the building structure on or in which they are to be placed, as required by the construction documents or governed by good practice.

The larger-sized conduits are rarely installed in reinforced concrete floor slabs, other than the ground floor slab, because of the possibility of weakening the building structure. It is necessary for the electrical inspector to fully understand these relationships to make sure that raceways are properly installed in accordance with NEC requirements and accepted practices in the trade.

When raceways are installed on the surface of building structures, hangers, supports or straps, and the necessary anchors, inserts, or other fastenings are required and must be provided as set forth in the NEC.

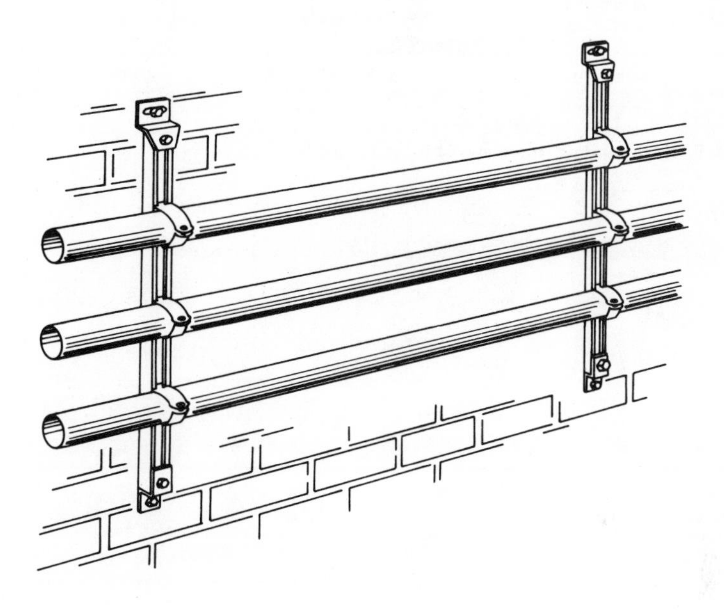

FIG. 4-7. When raceways are installed on the surface of building structures, they must be supported to comply with the NEC.

All conduit joints must be cut square, threaded where necessary, reamed smooth, and drawn up tight. Bends and offsets must be made with an approved bender or hickey and the number of bends per run must conform to the NEC limitations.

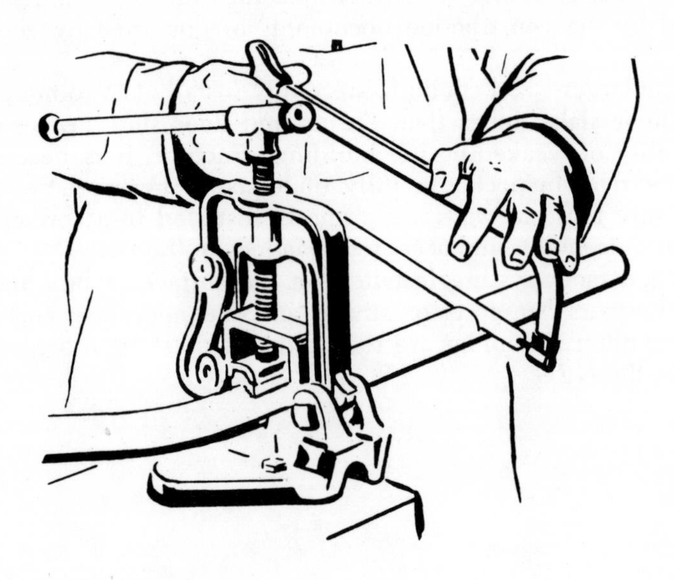

FIG. 4-8. All conduit joints must be cut squarely.

The conductors and equipment for delivering energy from an electrical supply system, such as the power company's lines, to the wiring system of a building or premises is called the (electric) service. Article 100 of the NEC gives definitions of the various parts of the electric service as follows:

Service Conductors The supply conductors that extend from the street main or from transformers to the service equipment of the premises supplied.

Service Drop The overhead service conductors from the last pole or other aerial support to and including the splices, if any, connecting to the service-entrance conductors at the building or other structure.

Service-Entrance Conductors, Overhead System The service conductors between the terminals of the service equipment and a point usually outside the building, clear of building walls, where joined by tap or splice to the service drop.

Service-Entrance Conductors, Underground System The service conductors between the terminals of the service equipment and the point of connection to the service lateral.

Service Lateral The underground service conductors between the street main, including any risers at a pole or other structure or from transformers, and the first point of connection to the service-entrance conductors in a terminal box or meter or other enclosure with adequate space, inside or outside the building wall. Where there is no terminal box, meter, or other enclosure with adequate space, the point of connection shall be considered to be the point of entrance of the service conductors into the building.

5–1(b). Certain NEC requirements are also noted on these drawings for quick reference by the electrical inspector.

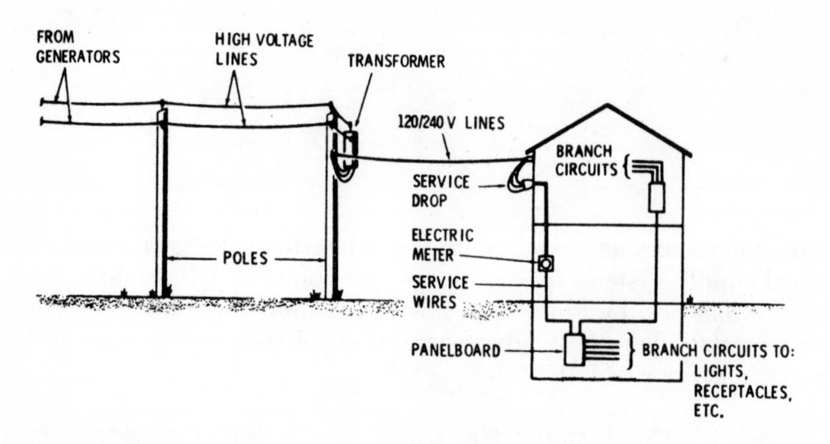

FIG. 5-1(a). Various sections of a typical overhead service entrance.

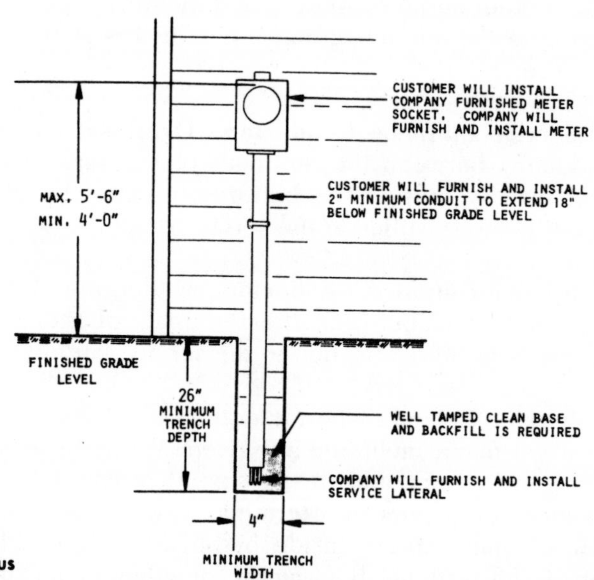

FIG. 5-1(b). Various parts of a typical underground service entrance.

5.1 TYPE AND SIZE OF SERVICE CONDUCTORS

Service conductors must be of a type to withstand exposure to atmospheric and other conditions of use without detrimental leakage of current. For service drops, all ungrounded conductors must be insulated (a grounded conductor may be bare). The conductors must be sized to sufficiently carry the calculated load, but in any case, shall not be smaller than No. 8 AWG copper or No. 6 aluminum.

For conductors used in underground services, it is recommended that all conductors be insulated. However, the NEC permits the grounded conductor in underground services to be bare when:

1. Used in a raceway.

2. Bare copper is judged to be suitable for the soil condition.

3. Bare copper or aluminum is used in an approved cable assembly having a moisture- and fungus-resistant outer covering.

Service-entrance conductors shall be insulated with the exception of the grounded conductor, which may be bare when used in a raceway or part of a service cable assembly, provided the cable terminates in an approved switch or circuit-breaker enclosure immediately upon entering the building.

Service-entrance cable (Type SE) is used quite frequently for residential services above ground, and Type USE is approved for

FIG. 5-2. Service-entrance cable.

underground use. This type of cable is also permitted for use as branch circuits or feeders if all conductors are insulated. If the grounded conductor is not insulated, Type SE cable may be used only as a branch circuit to supply a range, wall-mounted oven, counter-mounted cooking unit, or clothes dryer. Of course, this type of cable is permitted for interior use where the fully insulated conductors are used for circuit wiring and the uninsulated conductor is used for equipment grounding purposes.

Service-entrance cables used for commercial and industrial installations are normally enclosed in a raceway system. The conductors will have Type TW, THW, THWN or similar types of insulation. The electrical inspector should make certain that the proper size conductors are used; that is, the calculated load of the installation should be checked against Tables 310–16 through 310–19 of the NEC for proper size.

In general, no ungrounded conductors should be smaller than to carry 100 amperes in a 120/240 volt, 3 wire service. The grounded (neutral) conductor must be of sufficient size to carry the unbalanced load.

5.2 INSTALLATION OF SERVICE CONDUCTORS

Service-entrance conductors are limited to the following methods:

1. Open wiring on insulators
2. Rigid metal conduit.
3. Intermediate metal conduit
4. Electrical metallic tubing
5. Service-entrance cable
6. Wireways
7. Busways
8. Auxiliary gutters
9. Rigid nonmetallic conduit
10. Cablebus

11. Type MC cable

12. Mineral-insulated metal-sheathed cable

From the above list, it is evident that such cables as Type NM, AC, UF, and the like are not permitted. Furthermore, the twelve methods listed above have certain limitations and each method should be checked in Article 300 of the NEC.

Service-entrance conductors should not be spliced, nor should they be installed in a raceway with conductors other than service conductors. Both service raceways and cables must be equipped with a raintight service head. The service drop should not be readily accessible and must have the following clearances:

8 feet above the highest point of roofs over which they pass.

10 feet above finished grade, sidewalks, or from any platform or projection from which they might be reached.

12 feet over residential driveways and commercial areas such as parking lots and drive-in establishments not subject to truck traffic.

The restrictions on the use of particular wiring methods in particular situations are more critical for service conductors than in the case of feeders and branch circuits, as there is usually no overcurrent protection between the service entrance conductors and the power company source of energy supply of either overhead service drops or underground cables except the overcurrent protection on the transformer.

FIG. 5-3. Both service raceways and cables must be equipped with a rain tight service head.

5.2.1 Grounding of Services and Feeders

In general, the electrical wiring system is grounded at one point—at the service entrance location. The grounding should be accomplished by using a single conductor, which is connected at one end to a cold-water pipe or a driven ground rod, or both. The other end of the grounding conductor is connected to the service entrance metal raceway, or to the neutral bus of the main distribution panel.

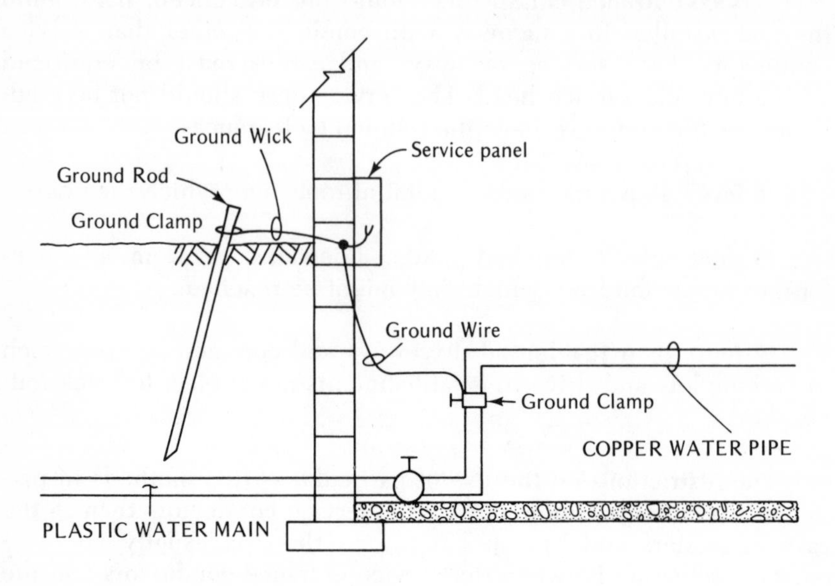

FIG. 5-4. One way of grounding a service entrance.

While the NEC considers the metallic enclosures of service entrance equipment to be adequately grounded when the service entrance raceway is mechanically connected to it, some local ordinances require that a grounding jumper conductor also be installed from the service raceway by means of a grounding bushing to the neutral bus in the main distribution panel. This arrangement is also sometimes required at the meter socket placed in the service raceway system.

The feeder metallic raceways and the metal enclosures of panelboards and distribution centers are considered to be adequately grounded when they are mechanically connected to each other and to the service entrance equipment metal enclosure.

Electrical lighting fixtures, wiring devices, and other electrical equipment are considered to be grounded when their metal parts

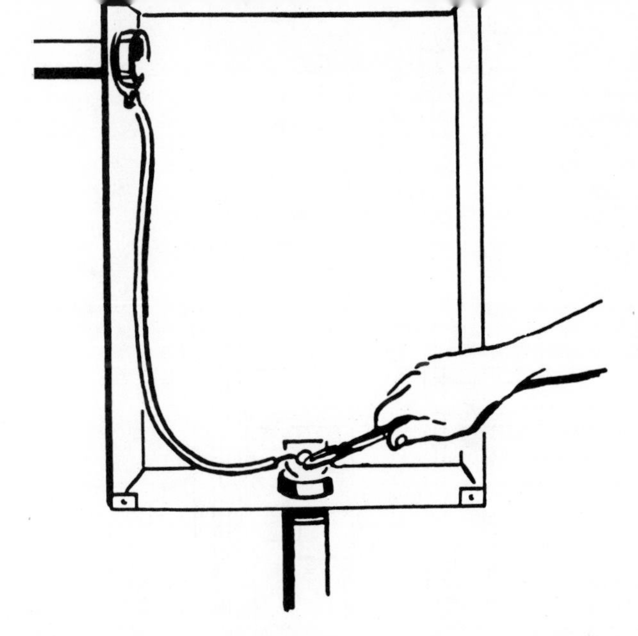

FIG. 5-5. Use of grounding bushings and jumper wire to ensure grounding continuity.

are connected to the outlet of a metallic branch circuit wiring system, either directly or by means of a grounding conductor incorporated in the appliance or equipment connecting cord.

In the case of nonmetallic branch circuit wiring systems, a copper grounding wire is connected to the outlet box and continues from outlet to outlet as a part of the cable assembly.

5.3 FEEDERS

Electric power is delivered to panelboard locations by feeder conductors extending from the main service entrance equipment to the branch circuit panelboards. The feeder conductors must be of sufficient size to meet the requirement of NEC Article 220, and are calculated roughly on a minimum basis of allowing for the connected lighting load or a certain number of watts per sq ft for lighting, plus the power equipment or appliances load requirements. In some cases, a demand factor is allowed for the connected power load.

On larger installations, distribution centers are established, to which point large feeder conductors are run from the main service

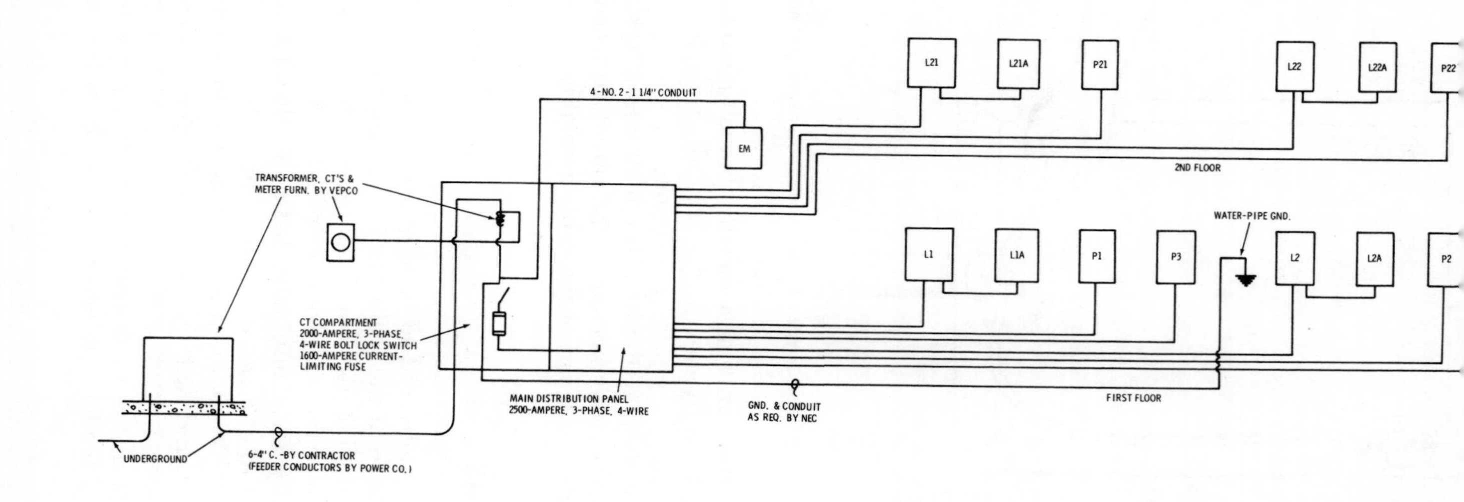

FIG. 5-6. Wiring diagram showing how electric power is delivered from the main distribution panel to branch circuit panelboards.

equipment. Subfeeders are then run from the distribution center to the various lighting and power panelboards.

When the service entrance voltage is higher than that required for the branch circuit wiring, step-down transformers are used near the panelboard locations to reduce the voltage of the conductors entering the panelboards. The usual voltages in such cases are 480/277 volts for the feeders and 120/240 volts for the branch-circuit panelboards.

FIG. 5-7. Step-down transformers are frequently used to reduce the voltage of feeders for use in branch circuit panelboards.

Type DT-3, Three Phase
45-1500 Kva, 5000 Volts and Below

In some instances, especially on industrial applications, higher voltage feeders, such as 2300 volts, are provided. The principle is the same as described in the previous paragraph, the difference being in the conductor insulation and the type of transformer and switching equipment.

The panelboards supplied by feeder conductors contain branch circuit overcurrent protective devices in the form of fuses, circuit breakers, and relays. They are usually grouped together at one or more centralized locations to keep the length of the branch circuit conductors to a practical minimum. These protective devices are usually factory-assembled and placed in a metal housing—the entire assembly being commonly called a panelboard or load center. The number and size of the protective devices vary in accordance with the branch circuit wiring system requirements. Circuit disconnecting switches are usually provided with fused panels.

In some cases, a power panel will be made up on the job site by assembling individually enclosed fused switches, circuit breakers, and relays or motor starter, together with a length of auxiliary gutter. The various components are connected by means of short conduit nipples or gutter connectors and insulated wires.

FIG. 5-8. A power panel installation using an assembly of individually enclosed fused switches.

Feeder raceways from the main distribution center to the branch circuit panelboards or heavy electrical equipment may be installed under concrete slabs; exposed on walls, ceilings, or beams and trusses; or concealed in spaces forming the finished ceilings, and in the walls. Vertical feeder runs are often installed in pipe shafts provided in the building structure. The main points for the electrical inspector's consideration are:

1. The proper wiring method must be used.

2. Raceways or cables must be secured as required in the NEC.

3. Conductors must be of the proper size and have the proper insulation for the use intended.

4. Terminations must be performed in a proper manner using correct connectors and other materials.

Chapter 6
Panelboards And Switchboards

A panelboard consists of a metal housing containing buses and automatic overcurrent protection devices, and are with or without switches for the control of lighting and power circuits. The housing or cabinet is installed against a wall or partition or can be supported by a frame. It is accessible only from the front. See Fig. 6-1.

A switchboard performs the same general function as a panelboard, but a switchboard consists of a large single panel, frame, or an assembly of panels on which are mounted (on the face or back or both) switches, overcurrent and other protective devices, buses, and sometimes measuring and testing instruments. See Fig. 6-2.

Safety switches fall into the category of panelboards since they are used in general electrical distribution and branch circuits and

FIG. 6-1. A panelboard consisting of automatic circuit breakers and a metal housing.

FIG. 6-2. Typical switchboard consisting of a large panel, frame, switches, over-current and other protective devices, buses, and measuring and testing instruments.

provide control and overcurrent protection. Safety switches are rated in amperes and horsepower (for use on motors) and each is capable of interrupting its rated current at its rated voltage.

6.1 CLASSIFICATION OF PANELBOARD HOUSINGS

Panelboard housings are manufactured in many different types for use in a number of different atmospheric conditions. Classification of standard types of enclosures for panelboards, motor starters, safety switches, etc., as designated by NEMA (National Electrical Manufacturers Association) is as follows:

Type 1 General Purpose A general-purpose enclosure is intended primarily to prevent accidental contact with the enclosed apparatus. It is suitable for general-purpose applications indoors where it is not exposed to unusual service conditions.

FIG. 6-3. A general-purpose enclosure.

Type 2 Driptight A driptight enclosure is intended to prevent accidental contact with the enclosed apparatus and, in addition, is so constructed as to exclude falling moisture or dirt.

Type 3 Weather-Resistant (Weatherproof) A weather-resistant enclosure is intended to provide suitable protection against specified weather hazards. It is suitable for use outdoors.

Type 4 Watertight This type of enclosure is suitable for application outdoors on ship docks and in dairies, breweries, etc.

FIG. 6-4. A weather-resistant enclosure.

Type 5 Dusttight A dusttight enclosure is provided with gaskets or their equivalent to exclude dust. It meets the requirements of dusttight.

Type 6 Submersible A Type 6 enclosure is suitable for application where the equipment may be subject to submersion, as in quarries, mines, and manholes. The design of the enclosure will depend upon the specified conditions of pressure and time.

Type 7 (A, B, C, or D) Hazardous Locations—Class I, Air Break These enclosures are designed to meet the application requirements of the National Electrical Code for Class I hazardous locations that may be in effect from time to time. In this type of equipment, the circuit interruption occurs in air. The letter or letters following the type number indicate the particular group or groups of hazardous locations (as defined in the National Electrical Code) for which the enclosure is designed. The designation is incomplete without a suffix letter or letters.

Type 8 (A, B, C, or D) Hazardous Locations—Class I, Oil-Immersed These enclosures are designed to meet the application requirements of the National Electrical Code for Class I hazardous locations that may be in effect from time to time. The apparatus is immersed in oil. The letter or letters following the type number indicate the particular group or groups of hazardous locations (as defined in the National Electrical Code) for which the enclosure is designed.

Type 9 (E, F, or G) Hazardous Locations—Class II These enclosures are designed to meet the application requirements of the National Electrical Code for Class II hazardous locations that may be in effect from time to time. The letter or letters following the type number indicate the particular group or groups of hazardous locations (as defined in the National Electrical Code) for which the enclosure is designed. The designation is incomplete without a suffix letter or letters.

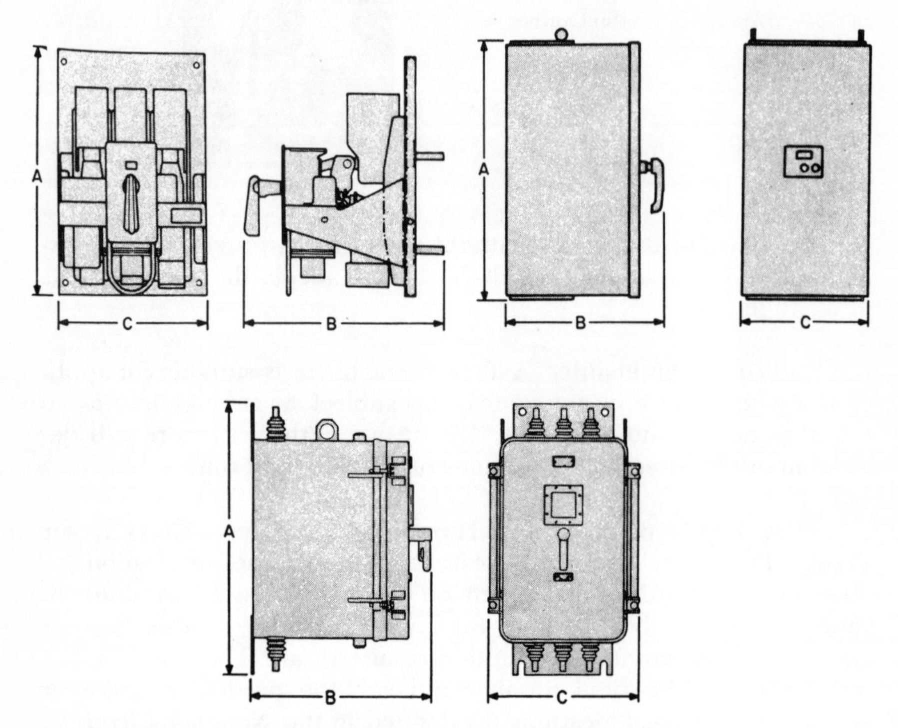

FIG. 6-5. A class I air break enclosure for hazardous locations.

Type 10 Bureau of Mines—Explosion-Proof A Type 10 enclosure is designed to meet the explosion-proof requirements of the U.S. Bureau of Mines that may be in effect from time to time. It is suitable for use in gassy coal mines.

Type 11 Acid- and Fume-Resistant—Oil-Immersed This enclosure provides for the immersion of the apparatus in oil such that it is suitable for application where the equipment is subject to acid or other corrosive fumes. A Type 11 enclosure is suitable for application indoors where the equipment may be subject to corrosive acid or fumes as in chemical plants, plating rooms, sewage plants, etc. The apparatus is immersed in oil.

Type 12 Industrial Use A Type 12 enclosure is designed for use in those industries where it is desired to exclude such materials as dust, lint, fibers, and flyings; and oil seepage or coolant seepage.

6.2 SAFETY SWITCHES

The electrical inspector should check all fusible safety switches to see that they are the quick-make, quick-break type (to eliminate arcing), with visible blades and dual horsepower rating. Handles on all switches must physically indicate ON and OFF positions. In most cases—especially on commercial and industrial installations—the switch handles should be able to accept padlocks having heavy duty industrial type shackles. The housing doors should be interlocked with the switch handles to prevent opening in the ON position.

Safety switches rated at over 1200 amperes at 250 volts or less, and at over 600 amperes over 250 volts, are intended for use only as isolating switches and must not be opened under load. To interrupt loads over this rating, a circuit breaker or a switch of special design approved for such purpose must be used (Article 380–13 NEC).

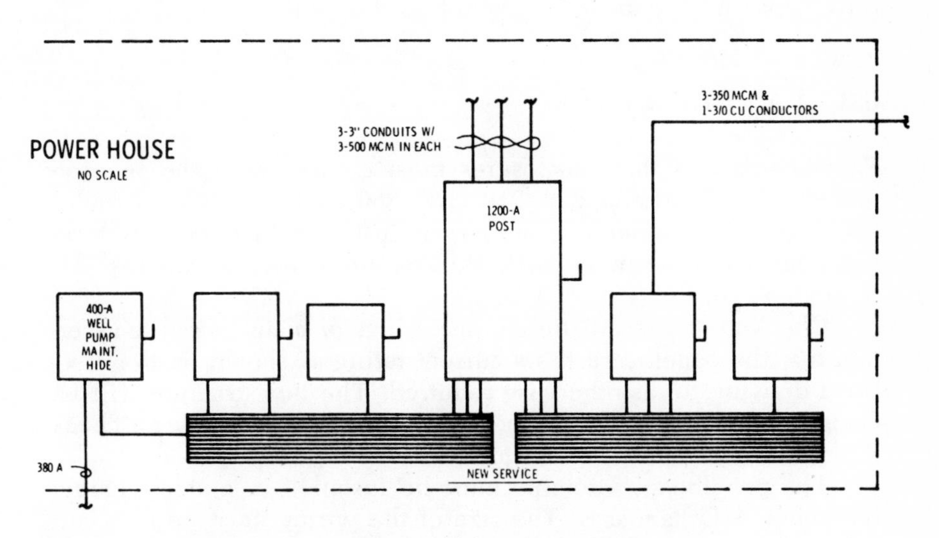

FIG. 6-6. A group of safety switches rated from 100 to 1200 amperes.

A safety switch must never be wired so as to disconnect the grounded conductor of a circuit except where the switch *simultaneously* disconnects *all* conductors of the circuit.

All safety switches must be installed where they can be operated from a readily accessible place. Furthermore, the center of the grip of the operating handle of the switch must not be mounted more than $6\frac{1}{2}$ ft above the floor or working platform. The three exceptions are as follows:

Exception No. 1 On busway installations, fused switches and circuit breakers shall be permitted to be located at the same level as the busway. Suitable means shall be provided to operate the handle of the device from the floor.

Exception No. 2 Switches installed adjacent to motors, appliances, or other equipment they supply shall be permitted to be located higher than specified in the foregoing and to be accessible by portable means.

Exception No. 3 Hookstick operable isolating switches shall be permitted at heights of more than 6 ft, 6 in.

As with all metal housings containing electrical components, the enclosure for safety switches of over 150 volts to ground must be grounded. Check Article 250 of the NEC. Where nonmetallic enclosures are used with metal-sheathed cables or metallic conduits, provisions must be made for grounding continuity.

6.3 PANELBOARDS

Panelboards and their enclosures must comply with the requirements of NEC Articles 240, 250, 370, 380, 384, and other Articles that apply. For example, panelboards installed in hazardous locations must also comply with the requirements of Articles 500 through 517 of the NEC.

Check the bus structure and main lugs or main switch to ascertain that the panelboard has a current rating as shown on the electrical drawings or as otherwise required. The bus structure will be designed to accommodate plug-in or bolted circuit breakers; check the drawings or specifications to see which type is required.

The panelboard assembly must be installed in a steel cabinet that meets UL standards. The size of the wiring must be in accordance with UL and NEC standards. For commercial applications, the cabinets are usually required to be equipped with a front door and have fully concealed, self-aligning trim clamps. The fronts should be full-finished steel with rust-inhibiting primer and baked enamel finish.

Terminals for feeder conductors to the panelboard mains and neutral system must be suitable for the type of conductor specified as well as the branch circuit conductor connected to the overcurrent devices.

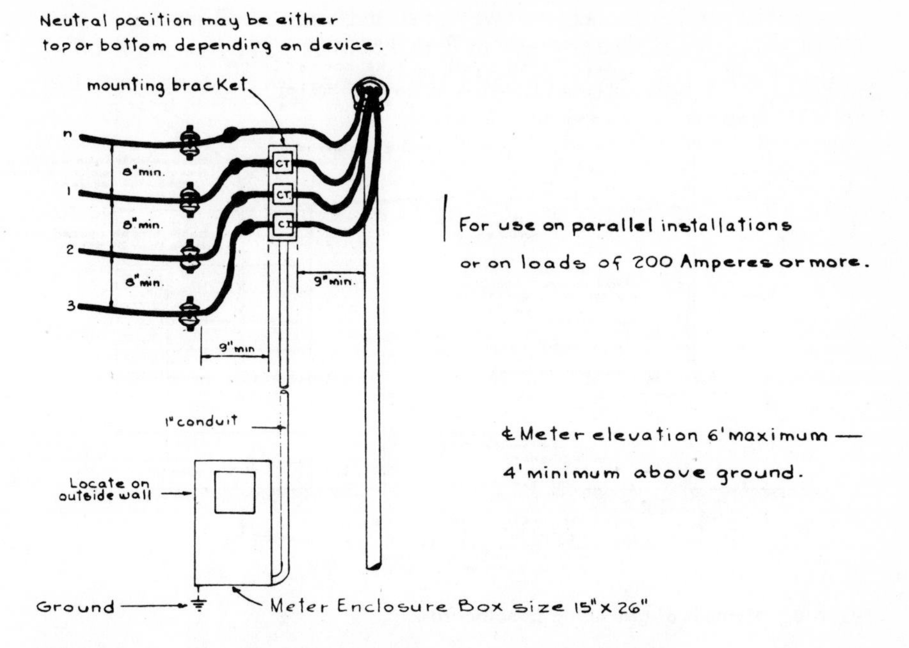

FIG. 6-7. Terminals for feeder conductors to the panelboard mains and neutral must be suitable for the type of conductor used.

Panelboards, conduit system attached to them, and the neutral system must be securely grounded according to Article 250 of the NEC. Once the ground has been established, tests should be carried out in the presence of the electrical inspector to verify that the ground has been established in accordance with NEC requirements and as specified in the construction documents. Written results of these tests should be kept on file in the inspector's office.

6.3.1 Branch Circuit Panelboards

In most cases, power and lighting panelboards should be of the dead front safety type with thermal magnetic, quick-make, quick-break, trip free, bolted- or plug-in type molded case circuit breakers. The voltage rating, number of poles, frame size, trip ratings, main breaker or lugs, neutral bus, and ground bus must be as shown on the drawings and in the written specifications and comply in every way with NEC requirements. Bus bars are usually rectangular, solid copper, and securely mounted and braced. Check to make certain that all connectors to the bus bars are securely bolted

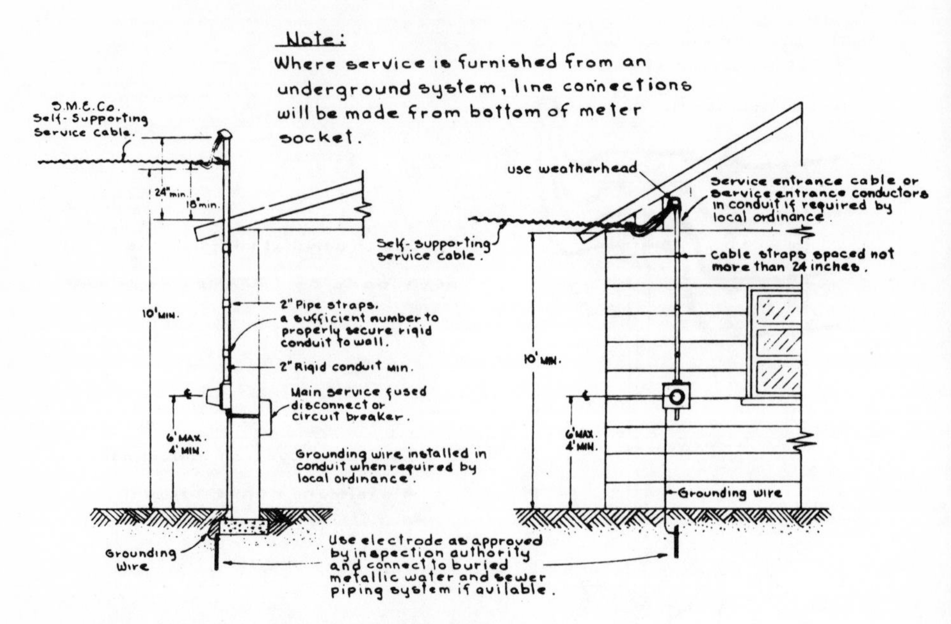

FIG. 6-8. Methods of grounding panelboards.

with the proper lugs for the size and type of conductors used. Many times, electricians will try to secure a larger conductor to the bus bar lugs than the lugs are designed to accept. This requires the cutting and removing of some of the strands of the conductors, thus lowering its current-carrying capacity. If such an act is suspected, the inspector should have the contractor remove the conductor from the lugs for visual inspection.

Cabinet boxes must be constructed of galvanized steel and/or constructed according to Article 373 of the NEC. In general, the cabinet boxes are required to be sized to provide a minimum of four-in. wide wiring gutters on sides, top, and bottom. The fronts should be constructed of code gauge steel, adjustable-indicating trim clamps, and with door provided with concealed hinges and cylinder lock and catch.

The panel directories should be neatly typed indicating the function of each circuit breaker in terminology approved by the electrical inspector.

According to Article 384–15 of the NEC, no more than 42 overcurrent devices (other than those provided for the mains) of a lighting or branch circuit panelboard shall be installed in any one cabinet or cutout box. Be aware that a 2-pole breaker is considered as two overcurrent devices; a 3-pole breaker three overcurrent devices, and so forth.

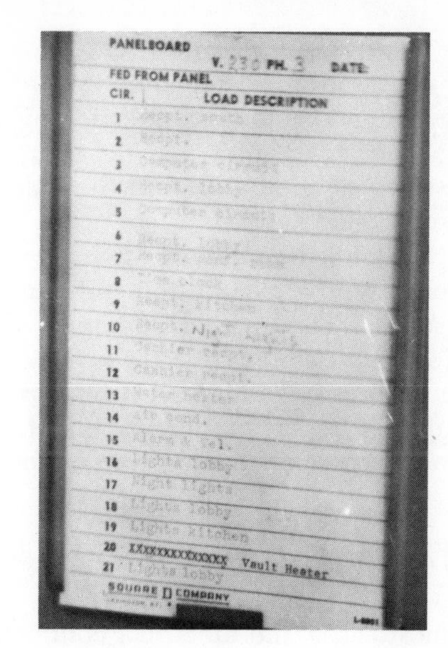

FIG. 6-9. Panel directories should be neatly typed indicating the function of each circuit breaker in terminology approved by the electrical inspector.

6.4 SWITCHBOARDS

Conductors and busbars on a switchboard must be located so they are free from damage and are firmly secured in place. Only those conductors intended for termination in a vertical section of a switchboard may be located in that section. Barriers must be placed in all service switchboards to isolate the service busbars and terminals from the remainder of the switchboard.

One common fault of switchboard installations is the arrangement of busbars and conductors so as to cause induction effects—causing the components to overheat. The electrical inspector should check the arrangement of all busbars and conductors on switchboards, and call attention to any faults; make certain that such faults are corrected prior to energizing the switchboard.

If the switchboard is used as service equipment, make certain that a main bonding jumper (sized in accordance with Section 250-79C of the NEC) or the equivalent is placed within the service disconnect section to be used for connecting the grounded service conductor on its supply side to the switchboard or panelboard frame. All sections of the switchboard must be bonded together using an equipment grounding conductor sized in accordance with Table 250-95 of the NEC.

Load terminals in switchboards must be arranged so that it is not necessary to reach across or beyond on ungrounded line bus in order to make load connection. Furthermore, where conduits or other raceways enter a switchboard at the bottom, sufficient space must be provided to permit the installation of conductors in the enclosure.

Switchboards having any exposed live parts must be located in a permanently dry location and must be accessible only to qualified personnel. The switchboard must also be under competent supervision at all times. For these reasons, switchboards with exposed live parts are seldom used in current electrical installations; they have been replaced with panels or switches that have all live electrical components enclosed in a metallic housing.

6.5 METERING

In most cases, the electrical contractor is responsible for furnishing and installing all raceways, current transformer cabinet, and meter trim for metering facilities for a given project. The local utility company will furnish the meter, current transformers, and necessary connections.

The electrical inspector should know the local utility company's requirements and make certain that the electrical contractor has complied before approving an electrical project. The electrical contractor is usually responsible for furnishing all miscellaneous components not provided by the utility company.

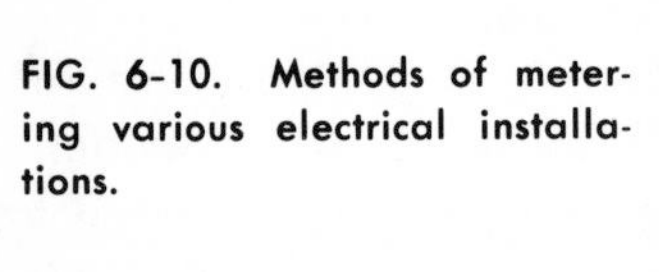

FIG. 6-10. Methods of metering various electrical installations.

QUICK REFERENCE CHECKLIST

Panelboards and Switches

_______ Enclosure approved for the location used.

_______ Electrical characteristics match the working drawings or conditions.

_______ Service or feeder conductors are of the proper size and type of panelboard supplied.

_______ Overcurrent protection adequate.

_______ Safety switches are of the quick-make, quick-break type and properly rated for the load.

_______ Adequate grounding.

_______ Tests made for proper grounding.

_______ Properly fused for the load and circuit.

_______ Panelboards and switches readily accessible.

_______ Location of panelboards and switches of correct height so that circuit breaker or switch handle is not more than 6½ feet above the floor or working platform.

_______ Circuit breakers of correct size, poles, and frame.

Chapter 7
Lighting Fixtures

Every building should contain electric lighting that will provide the highest visual comfort and performance consistent with the type of area to be lighted and the budget provided. To achieve this goal, architectural and engineering firms have lighting designers calculate the illumination requirements for given areas and then select lighting fixtures that will provide the illumination obtained in the calculations. The electrical inspector should therefore examine the drawings and specifications carefully and then check the construction documents against the finished installation.

Most designers prefer to specify lighting equipment by exact description (manufacturer, catalog number, etc.) while others specify the equipment by performance, giving the contractor an option of several brands. Since basic types of lighting fixtures vary widely

LIGHTING FIXTURE SCHEDULE						
FIXT TYPE	MANUFACTURER'S DESCRIPTION	LAMPS		V	MOUNTING	REMARKS
		NO.	TYPE			
1	BENJAMING CAT NO. AG-7244-4	4	40W F	120	RECESSED	
2	CRESCENT CAT NO. ANG 220	2	20W F		WALL	
3	MOLDCAST CAT NO. 2100	1	150WI		WALL	
4	BENJAMIN CAT NO. 9642	2	150WI		SURFACE	
5	BENJAMIN CAT NO DC-2214-4	1	40W F		SURFACE	
6	MOLDCAST CAT NO 531	1	150WI		WALL	
7	MOLDCAST CAT NO. A-270	1	150WI			
8						
9						
10						

FIG. 7-1. Most lighting designers prefer to specify lighting equipment by exact description as shown in this lighting fixture schedule.

in quality, design and physical appearance, mechanical features, brightness, and finishes, it is difficult to provide the electrical inspector with concrete data to check a specific installation. However, the following paragraphs will cover the general inspection procedures for the majority of lighting installations.

7.1 LIGHTING CALCULATIONS

Table 220-2(b) of the NEC gives general lighting loads of several occupancies. A unit load of not less than those specified should be provided. The floor area for each floor should be computed from the outside dimensions of the building and for dwelling units, the computed floor area need not include open porches, garages, or unused or unfinished spaces not adaptable for future use. Table 220-2(b) reads as follows:

TABLE 220-2(B)

Type of Occupancy	Unit Load per Sq Ft (watts)
Armories and auditoriums	1
Banks	5
Barber shops and beauty parlors	3
Churches	1
Clubs	2
Court rooms	3
Dwelling units	3
Garages—commercial (storage)	½
Hospitals	2
Hotels and motels	2
Industrial commercial buildings	2
Lodge rooms	5
Office buildings	5
Restaurants	2
Schools	3
Stores	3
Warehouses	¼

Diversity factors may be applied to some of the occupancies listed above. In applying demand factors they shall apply to that portion of the total branch-circuit load computed for general illumination. See Table 220-11 of the NEC.

TABLE 220-11

Type of Occupancy	Load to Which Demand Portion of Lighting Load to Which Demand Factor Applies	Demand Factor Demand Factor Percent
Dwelling Units	First 3000 or less at	100
	Next 3001 to 120,000 at	35
	Remainder over 120,000 at	25
Hospitals	First 50,000 at	40
	Remainder over 50,000 at	20
Hotels and motels	First 20,000 at	50
	Next 20,001 to 100,000 at	40
	Remainder over 100,000 at	30
Warehouses	First 12,500 at	100
	Remainder over 12,500 at	50
All others	Total wattage	100

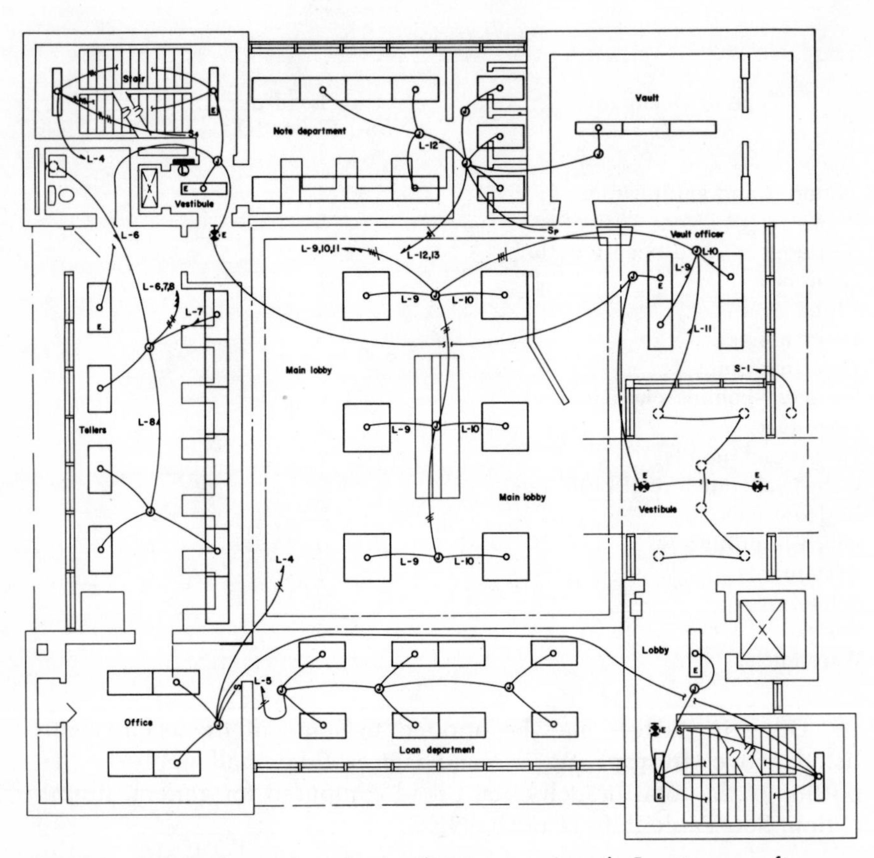

FIG. 7-2. Lighting layout for a bank utilizing approximately 5 watts per sq ft.

For show-window lighting, a load of not less than 200 watts shall be included for each linear foot of show window, measured horizontally along its base.

7.2 OUTLETS REQUIRED

The NEC has no set provisions for the exact location of most lighting fixtures in various areas. However, for residential occupancies, Article 210-26 requires that at least one wall switch controlled lighting outlet be installed in every habitable room, in bathrooms, hallways, stairways, and attached garages, and at all outdoor entrances. At least one lighting outlet must be installed in an attic, underfloor space, utility room, and basement only where these spaces are used for storage or containing equipment requiring servicing.

Following are examples of how the requirements of Article 210-26 are met: A home containing a crawl space under the house with no access door or equipment would not require a lighting fixture. However, if say, a small water pump and pressure tank were installed under the house in the crawl space, or a suitable door were provided so that the space could be used for storage, then a light would be required.

The only other NEC requirement concerning the placement of lighting outlets is that at least one wall switch controlled lighting outlet or wall switch controlled receptacle must be installed in guest rooms in hotels, motels, or similar occupancies.

7.3 EMERGENCY LIGHTING SYSTEMS

Municipal, state, federal, and other codes require that certain occupancies be provided with emergency lighting in the event of failure of the normal electric service. This is to insure the safety of life and property. These systems are normally installed in places of assembly where artificial illumination is required, that is, hotels, theaters, sports arenas, hospitals, and similar institutions.

When such systems are installed in buildings under the electrical inspector's jurisdiction, he should conduct or witness a test on the complete system upon installation and periodically afterward. The periodic tests are performed to assure that their maintenance keeps the emergency lighting systems in proper operating condition.

In general, an emergency lighting system should have adequate capacity and rating for the emergency operation of all lighting equipment connected to the system. The emergency power system may be supplied by either storage batteries or a generator set.

Individual emergency illumination units shall consist of the following:

1. a rechargeable battery;

2. a battery charging means;

3. provisions for one or more lamps mounted on the equipment and/or terminals for remote lamps;

4. a relaying device arranged to energize the lamps automatically upon failure of the supply to the unit equipment.

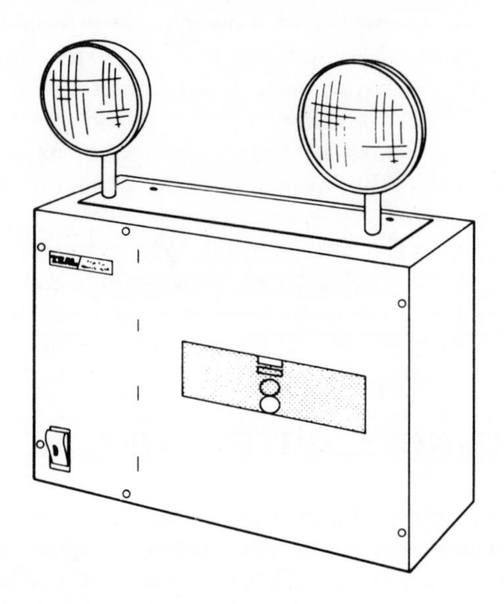

FIG. 7-3. Individual emergency illumination unit.

Emergency illumination must include all required exit lights and all other lights specified as necessary to provide sufficient illumination. Emergency lighting systems must be so designed and installed that the failure of any individual lighting element, such as the burning out of a light bulb, cannot leave any space in total darkness.

For branch circuits supplying emergency lighting, there should be an emergency supply source to which the load will be transferred automatically and immediately upon the failure of the normal supply. All overcurrent devices protecting emergency circuits should be accessible to authorized persons only. Furthermore, emergency circuit wiring must be kept entirely independent of all other wiring.

FIG. 7-4. Emergency illumination must include all required exit lights.

7.4 LIGHTING FIXTURES

Lighting fixtures—sometimes called luminaires—are complete lighting units consisting of a lamp or lamps together with the parts designed to distribute the light, to position and protect the lamps, and to connect the lamps to a power supply. Lighting fixtures range from a simple porcelain lampholder fastened directly to an outlet box to large and elaborate fixtures suspended by cables or supported on lighting standards or towers. They are usually classed as residential, commercial, industrial, decorative, or floor lighting.

The types of lamps include incandescent filament, fluorescent, high-intensity discharge, short-arc, miscellaneous discharge, electroluminescent, light emitting diodes, and carbon-arc. Lamps, other than incandescent, usually require auxiliary equipment such as ballasts or transformers for their operation, which are placed either in the fixture itself, adjacent to the fixture, or in an assembly of such equipment at central points remote from the fixture.

Lighting fixtures are installed in many ways and in many different configurations. For example, one type of lighting fixture is designed for surface mounting (see Fig. 7-5), another type is suspended (Fig. 7-6), while still another type is recessed so that the surface of the fixture is flush with the surface of the ceiling (see Fig. 7-7). Any of these fixtures may be installed as individual units or grouped to form continuous rows.

FIG. 7-5. Typical surface mounted lighting fixture.

FIG. 7-6. Lighting fixture suspended from ceiling.

FIG. 7-7. Flush mounted lighting fixture.

Luminous ceilings are being used extensively for illuminating residential kitchens as well as high-quality commercial buildings. This arrangement provides a "skylight" effect, which greatly reduces eye strain and makes seeing easier. A dimming control is sometimes added for even greater versatility.

A unique illuminated ceiling system with advantages not normally possible with regular systems is shown in Fig. 7-8. Because of the indirect lighting effect, it is possible to achieve a very evenly lighted ceiling, using a very shallow plenum space from 6 to 12 in. A sturdy extruded aluminum channel acts as the main support track with standard fluorescent strip fixtures mounted in an inverted position inside the main track, subsequently suspended a short distance below a standard acoustical or plaster ceiling.

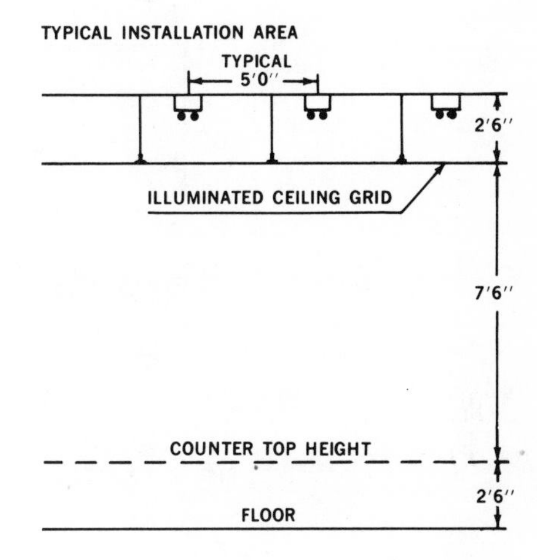

FIG. 7-8. Details of an illuminated ceiling system.

Standard lamp spacing of 28, 34, 40, and 52 in. gives lighting levels up to 75 footcandles using continuously mounted single-lamp 430-MA fixtures.

Fluorescent or incandescent lighting fixtures are also used behind horizontal or vertical coves that conceal the fixtures and provide indirect illumination for an area. This type of arrangement is ideal for modernizing old areas, as the coves can also be used to conceal the wiring. Vertical coves are normally referred to as valance or cornice lighting. Cornice lighting directs all light downward to give dramatic interest to wall coverings, draperies, and the like. See Fig. 7-9. Valances provide up-light, which reflects off the ceiling for general room lighting, and also down-light, for drapery or wall accent. Another use for valance lighting is over desks as shown in Fig. 7-10. Here the two 40-watt fluorescent lamps provide nearly

100 footcandles of illumination on the desk work area, and the fixtures also direct a portion of the light upward to illuminate the books on the shelf above the lighting fixtures.

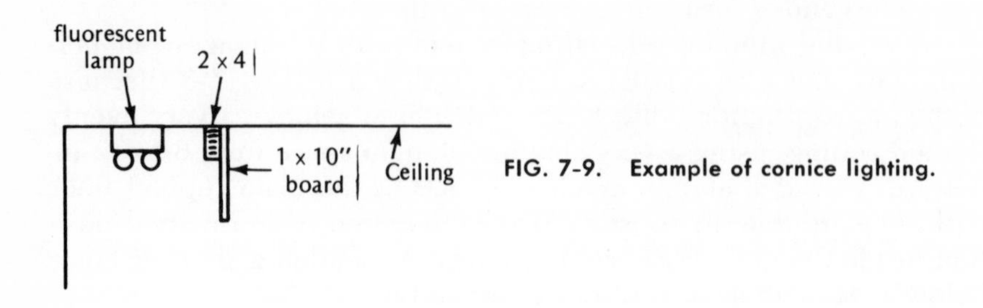

FIG. 7-9. Example of cornice lighting.

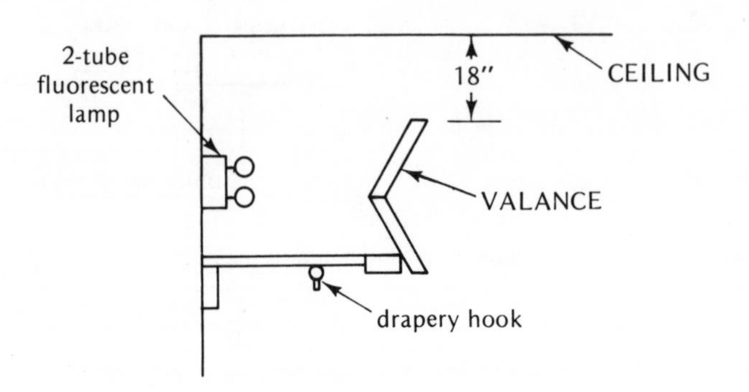

FIG. 7-10. Typical valance lighting.

Floodlighting fixtures of various types and sizes are being used increasingly in area lighting of storage yards, parking lots, recreation facilities, highways, the exterior of buildings, and fence security systems. Some examples are shown in Fig. 7-11.

Individual lighting standards with one or more luminaires of different types of light sources are used for street and highway lighting. Special lighting fixtures are widely used at intersections for traffic control and to give directions. Other types of specialized lighting fixtures have been developed for airport runway and beacon lighting.

It is not within the scope of this chapter to cover all necessary details of lighting design and application. However, the electrical inspector should inform himself of the basic principles of illumination. One book by the author titled *Principles of Illumination* is

FIG. 7-11. Example of flood light.

available from Howard W. Sams & Co., Inc., 4300 West 62nd Street, Indianapolis, Indiana 46206. Manufacturers of lighting fixtures and lamps have available a great amount of literature and catalog data that can be used to advantage by the electrical inspector in familiarizing himself with all the various items he must know to properly inspect lighting installations.

Chapter 8
Communication Equipment

The installation of systems falling under the heading of communication and alarm systems employs a wide variety of techniques, often involving special types of equipment and materials designed for specific applications. Many systems operate on low-voltage circuits but are installed similarly to conventional electrical circuits, that is, those used for light and power. In all cases, the installations must conform to applicable NEC requirements (especially those covered in Chapters 6 and 7 of the NEC), local ordinances, and to instructions provided by security, alarm and communication equipment manufacturers.

The items covered in this chapter encompass radio, shortwave, and microwave transmission; alarm and detection systems, smoke detectors; clock and program equipment; television systems; telephone and telegraph; intercommunication equipment; public address systems; and systems related to learning laboratories.

8.1 FIRE ALARM AND DETECTION SYSTEMS

Besides NEC requirements, fire alarm systems installed in buildings must meet the requirements of NFPA (National Fire Protection Association) Code 72, and all equipment should be UL listed.

The electrical inspector should obtain a copy of the manufacturer's specifications and installation instructions and then make certain that the system is actually installed according to these specifications. Upon completion of the project, the system should be in first class operating condition and should be tested by or in the presence of the electrical inspector.

All wiring for the fire alarm system should be in strict accordance with the NEC, NFPA Codes 70 and 72, and all local electrical codes applying. Check the number and size of conductors against the wiring diagram supplied by the manufacturer. These circuits must be identified at terminal and junction box locations in a manner which will prevent unintentional interference with the signaling circuit during testing and servicing. Those circuits extending beyond one building (aerially) must meet the requirements of Article 800 of the NEC and be classified as communications circuits, or must meet the requirements of NEC Article 225.

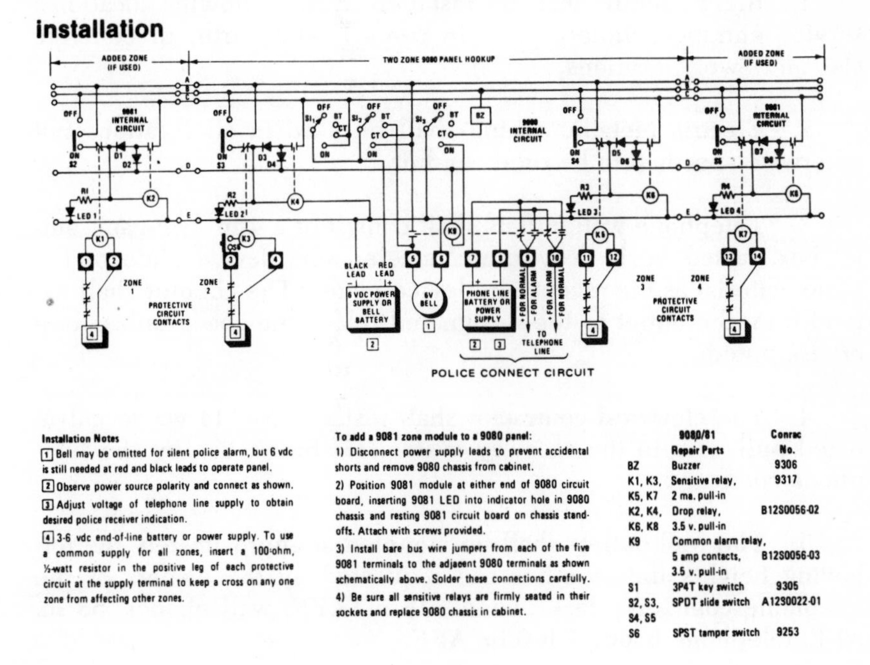

Installation Notes

[1] Bell may be omitted for silent police alarm, but 6 vdc is still needed at red and black leads to operate panel.

[2] Observe power source polarity and connect as shown.

[3] Adjust voltage of telephone line supply to obtain desired police receiver indication.

[4] 3-6 vdc end-of-line battery or power supply. To use a common supply for all zones, insert a 100-ohm, ½-watt resistor in the positive leg of each protective circuit at the supply terminal to keep a cross on any one zone from affecting other zones.

To add a 9081 zone module to a 9080 panel:

1) Disconnect power supply leads to prevent accidental shorts and remove 9080 chassis from cabinet.

2) Position 9081 module at either end of 9080 circuit board, inserting 9081 LED into indicator hole in 9080 chassis and resting 9081 circuit board on chassis standoffs. Attach with screws provided.

3) Install bare bus wire jumpers from each of the five 9081 terminals to the adjacent 9080 terminals as shown schematically above. Solder these connections carefully.

4) Be sure all sensitive relays are firmly seated in their sockets and replace 9080 chassis in cabinet.

	9080/81 Repair Parts	Conrac No.
BZ	Buzzer	9306
K1, K3, K5, K7	Sensitive relay, 2 ma. pull-in	9317
K2, K4, K6, K8	Drop relay, 3.5 v. pull-in	B12S0056-02
K9	Common alarm relay, 5 amp contacts, 3.5 v. pull-in	B12S0056-03
S1	3P4T key switch	9305
S2, S3, S4, S5	SPDT slide switch	A12S0022-01
S6	SPST tamper switch	9253

FIG. 8-1. Wiring diagram of a fire security system.

Fire alarm signaling circuits and equipment must be grounded in accordance with NEC Article 250, and all circuits must be electrically supervised so that a trouble signal will indicate the occurrence of a single open or a single ground fault on any installation wiring circuit that would prevent proper alarm operation. Check NEC Article 760 for further requirements.

8.2 TELEPHONE SYSTEMS

Although public telephone systems are normally installed by the local telephone company, such systems require that certain work be performed by the electrical contractor prior to the connection of the lines and telephones. Such work will include raceways, outlet boxes, and a pull wire inside the conduit or raceway. For all projects, the electrical inspector should check with the local telephone company to ascertain just what is required of the contractor and the owners. The requirements of one telephone company are as follows:

1. Rigid conduit shall be installed in the following locations: service entrance, underground in contact with earth, in concrete slab and "wet" locations.

2. Electric metallic tubing shall be used in all locations not otherwise specified to be rigid conduit.

3. Telephone wall outlets shall consist of a 4-in. two-gang outlet box, raised device cover, and a telephone device plate of the same material as the receptacle device plates. The conduit shall extend from the outlet to the designated telephone space unless otherwise noted.

4. The electrical contractor shall install a No. 14 gauge galvanized pull wire in the raceway system for future use by the telephone company.

5. The wall outlets shall be mounted at approximately the following heights unless otherwise requested by the owners or the telephone company: desk phones, 18 in. AFF; wall phones, 58 in. AFF; telephone booth, 7 ft 6 in. AFF.

When telephone lines are installed by the electrical contractor or when they fall under the jurisdiction of the electrical inspector, a protector approved for the purpose must be installed on each circuit run partly or entirely in aerial wire or aerial cable not confined within a block. Also, a protector approved for the purpose must be provided on each circuit, aerial or underground, so located within the block containing the building served as to be exposed to accidental contact with light or power conductors operating at over 300 volts to ground.

The protector should be located in, on, or immediately adjacent to the structure or building served and as close as practical to the point at which the exposed conductors enter or attach. The protector should not be located in the vicinity of easily ignitible material nor in any area classified as hazardous as defined in NEC Article 500.

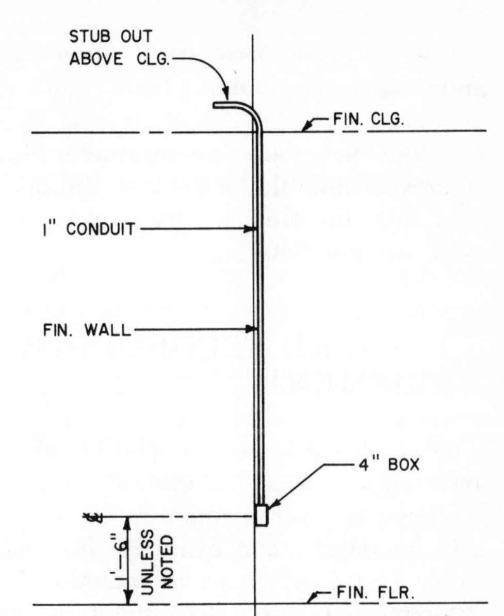

FIG. 8-2. Detail of telephone outlet and related raceway provided so that telephone wires can be pulled to the box from the accessible ceiling space.

Communication conductors must not be installed in any raceway, compartment, outlet box, junction box, or similar fitting with conductors of light or power circuits, except where the conductors of the different systems are separated by a suitable partition.

8.2.1 Grounding

Telephone and communication equipment must be grounded in an approved manner. Where the cable is exposed to contact with electric light or power conductors, the metal sheath of aerial cables entering buildings must be grounded or shall be interrupted close to the entrance to the building by an insulating joint or equivalent device. The protector shall comply with the following requirements:

1. The grounding conductor shall have a 30-mil rubber insulation and shall be covered by a substantial fibrous covering. Conductors approved for the purpose having less than 30-mil insulation or having other kinds of insulation shall be permitted.

2. The grounding conductor shall not be smaller than No. 18 AWG copper or equivalent.

3. The grounding conductor shall be run to the grounding electrode in as straight a line as practicable.

4. Where necessary, the grounding conductor shall be guarded from damage.

5. The grounding conductor shall be connected either to the nearest accessible location on the building grounding electrode system, or to the electric service grounded conductor, or as provided in NEC Section 800-31.

8.3 SOUND RECORDING AND SIMILAR EQUIPMENT

The electrical inspector who checks electrical installations in such buildings as theaters, schools, hospitals, hotels, apartment houses, commercial establishments, office buildings, and industrial plants will be required to evaluate the installation of various sound systems. These systems will include sound recording, public address, intercommunicating telephone, nurses' call systems. TV receiver and speaker, and other audible signal equipment and wiring.

In most instances, the type of stem will be detailed in the working drawings and specifications by a consulting engineering firm commissioned by the architect. However, many of these systems are installed by the electrical contractor without complete working drawings. In such cases, the electrical inspector should obtain installation instructions from the manufacturer of the equipment or from the electrical contractor installing the system.

Most of the circuits on sound systems will be low-voltage with the exception of the power supply. Conductors of varying sizes and types of insulation are used and either installed in raceways or supported by approved cable straps. Special outlet boxes or cabinets are usually provided with the equipment although some items may be mounted in or on standard outlet boxes.

Some systems require direct current, which is usually obtained by electronic rectifying equipment. The low-voltage alternating current is obtained through the use of relatively small transformers connected on the primary side to a 120-volt power source.

8.4 COMMUNICATION CIRCUITS

Central station communication circuits are installed overhead on system poles, or jointly with power distribution systems on jointly owned poles, or underground. Protective devices are required in communication circuits entering buildings to prevent the higher voltage power current from passing into the building in case of accidental contact of the two systems. When installed in raceways, communication circuits must be installed in separate raceways from the power circuit and, except when introduced for the sole purpose of a power supply, power circuits must not enter the same outlet boxes used for communication circuits unless they are separated by a partition.

When communication conductors are installed overhead, proper clearances from the power circuits must be maintained as well as clearances over streets, roads, and driveways. See Chapter 11 in this book.

Communication system conductors vary in size, type of insulation, and combination of conductors, depending upon the system requirements. They are installed as single conductors, but most often as nonmetallic sheathed cables often containing a large number of color-coded conductors.

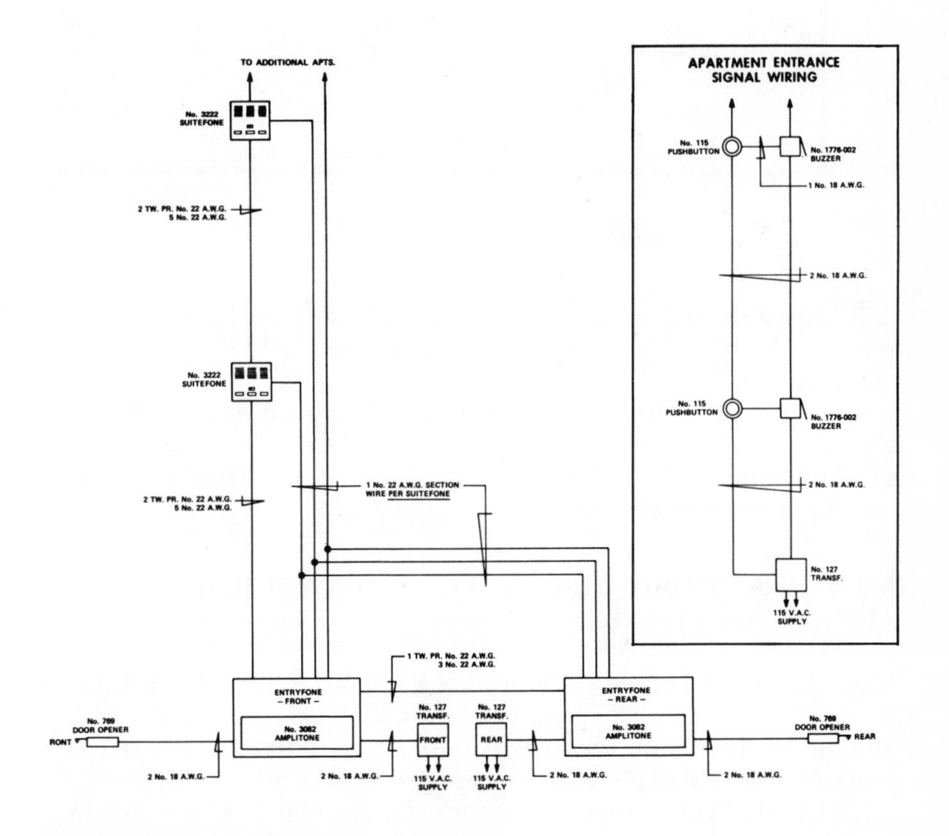

FIG. 8-3. Apartment house intercom system.

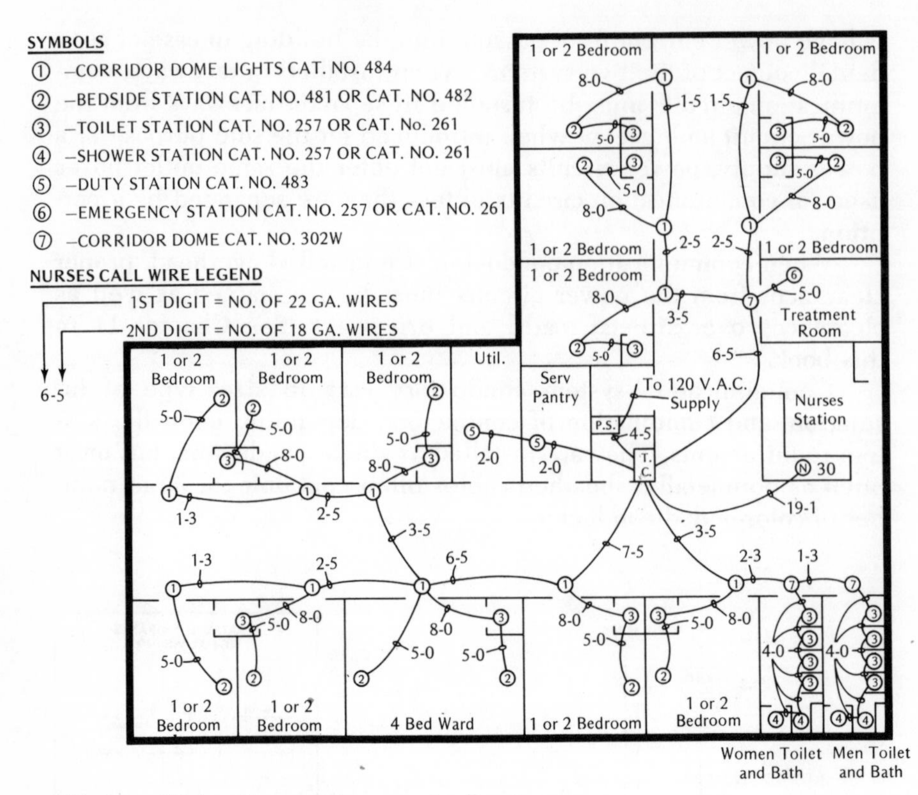

FIG. 8-4. Wiring diagram of a nurses' call system.

Conductors are connected either directly to the equipment or to terminal blocks. Nonmetallic or lead-sheathed cables are spliced when necessary to make long continuous runs, and are usually terminated in special terminal boxes to prevent moisture from entering the insulation. Lead-sheathed cables are installed overhead suspended from steel messenger cables, as well as installed underground in raceway systems.

8.4.1 Community Antenna Television and Radio Distribution Systems

Most community antenna systems use coaxial cable for the distribution of the radio and TV frequency signals. Article 820 of the NEC applies to these systems. When the wiring system employed is other than coaxial, Article 800 of the NEC shall apply.

The NEC permits the use of coaxial cable to deliver low-energy power to equipment directly associated with antenna (radio fre-

WIRING DIAGRAM
SYSTEM TE

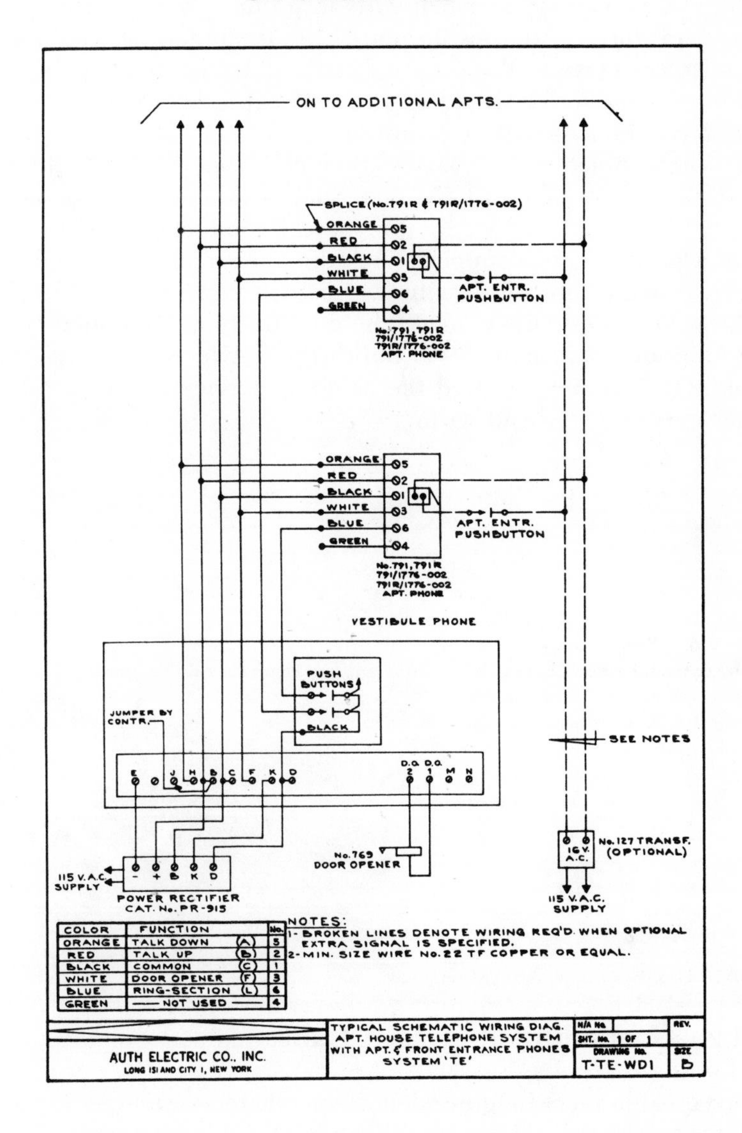

COLOR	FUNCTION	No.
ORANGE	TALK DOWN (A)	5
RED	TALK UP (B)	2
BLACK	COMMON (C)	1
WHITE	DOOR OPENER (F)	3
BLUE	RING-SECTION (L)	6
GREEN	—— NOT USED ——	4

FIG. 8-5. Wiring diagram of an intercom system.

quency) distribution system provided the voltage does not exceed 60 volts, and if the current supply is from a transformer or other device having energy-limiting characteristics.

Where the system is exposed to lightning or to accidental contact with lightning arrestor conductors or power conductors operating at a voltage of over 300 volts to ground, the outer conductive shield of the coaxial cable must be grounded at the building premises as close to the point of cable entry as practicable. Where the outer conductive shield of a coaxial cable is grounded, no other protective devices will be required.

Where practicable, coaxial conductors on poles should be located below the light or power conductors and must not be attached to a crossarm that carrries light or power conductors. Lead-in cables from a pole or other support, including the point of initial attachment to a building or structure, must be kept away from electric light or power circuits to avoid the possibility of accidental contact.

Conductors installed over buildings, between buildings, and on buildings must be installed the same as service entrance cables or other outside, overhead wiring.

FIG. 8-6. Power company pole with communication cable attached. Note the distance between the lower communications cable and the upper power line.

8.4.2 Grounding of Coaxial Cable

Coaxial cable must be grounded as specified below:

1. The grounding conductor must have a rubber or other suitable kind of insulation.

2. The grounding material must be copper or other corrosion-resistant conductive material, stranded or solid.

3. The grounding conductor must not be smaller than No. 18 AWG; it must have an ampacity approximately equal to that of the outer conductor of the coaxial cable.

4. The grounding conductor must be run to the grounding electrode in as straight a line as practicable.

5. Where necessary, the grouding conductor must be guarded from damage.

6. The grounding conductor must be connected either to an available cold water pipe, to the power service conduit, or to other means as described in Article 820-22 of the NEC.

Chapter 9
Electrical Equipment

Items covered under this chapter will include electric heating, electric motors, air-conditioning and refrigerating equipment, generators, transformers, capacitors, reactors, and storage batteries. The equipment and their related circuits are covered in Chapter 4 of the NEC, and the electrical inspector should refer to this section of the code frequently when such items are encountered on electrical construction work.

9.1 ELECTRIC HEATING

Besides space heating equipment consisting of electric baseboard heaters, fan (forced air) heaters, duct heaters, heating cable, and the like, the electrical inspector will encounter other electric heating devices such as snow melting cable and mats, roof and gutter heating cable, and packaged room heating and cooling units.

In general, the electric inspector should check all circuits feeding the equipment to insure that the proper size and type of conductors are used and especially that all connections to the heater junction box are made with an approved type of connector.

Except in certain special cases, all electric heating equipment should be automatically controlled by thermostats installed where indicated by the designer or in some cases built into the individual units. Thermostats provided with a marked OFF position are permitted to serve as both controllers and disconnecting means provided all of the following conditions are met:

1. The thermostat directly opens all ungrounded conductors when manually placed in the OFF position.

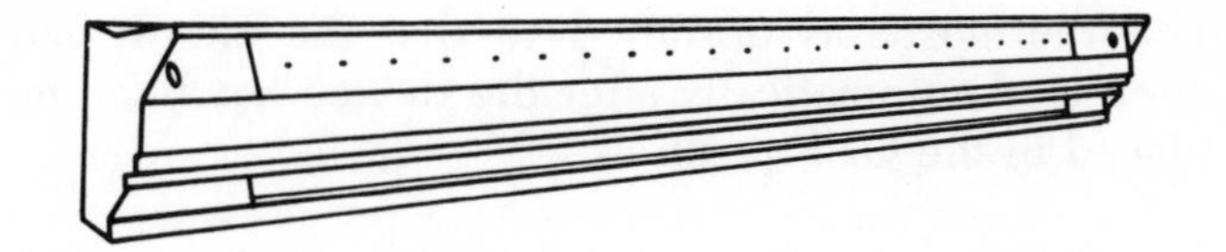

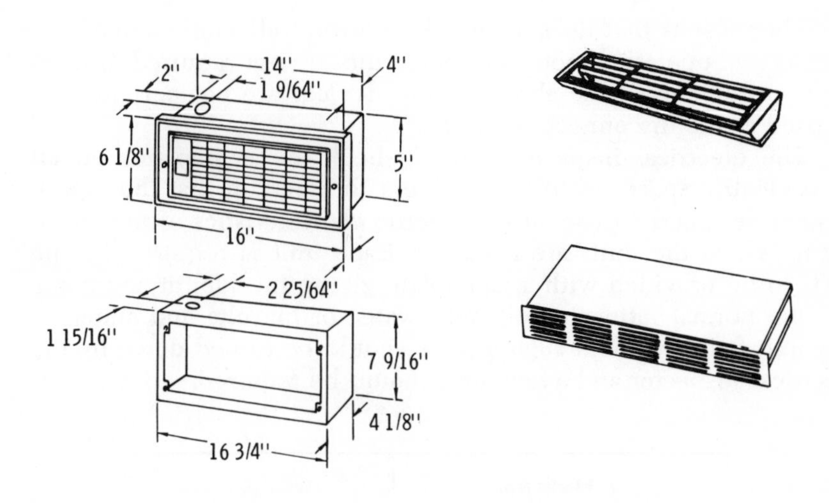

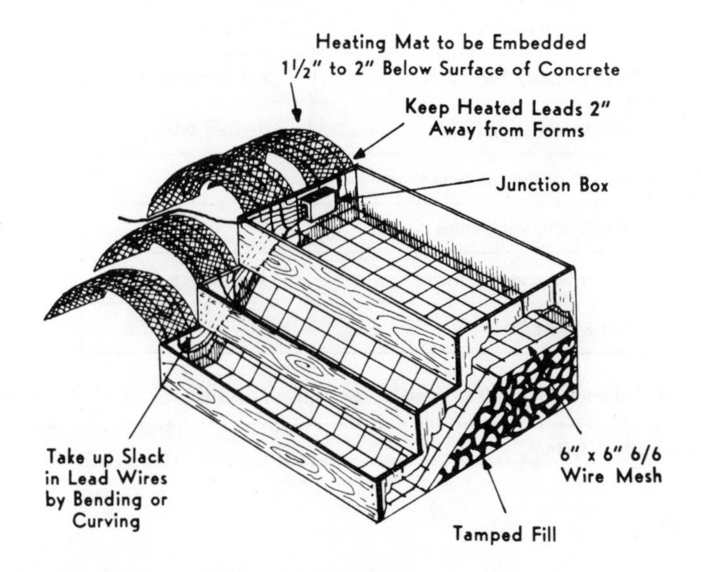

FIG. 9-1. Several types of electric heating equipment.

2. The thermostat is designed so that the circuit cannot be energized automatically after the device has been manually placed in the OFF position.

3. The thermostat is located as specified in Section 424–19 of the NEC.

Thermostats that do not directly interrupt all ungrounded conductors and operate remote control circuits are not required to meet the requirements of the above. These devices are not permitted to be used as the disconnecting means.

The electrical inspector should check the nameplates on all fixed electric space heating equipment to ascertain that the equipment is designed for use on the electric characteristics of the building in which the units are installed. Each unit is required by the NEC to be provided with a nameplate giving the identifying name and the normal rating in volts and watts, or in volts and amperes. Any unit found without such a plate should be turned down by the electrical inspector and a new unit should be required.

Model No.	WF6-15
Media size (in.)	9-1/2 x 26-1/2 x 1/2
Water supply connection (in.)	1/4 compression
Overflow fitting (in.)	3/8 o.d.
Inlet & outlet opening (diam. in.)	6
Motor operating amps	.15
Electrical characteristics	115V - 60cy - 1ph
Net weight (lbs)	17

FIG. 9-2. The electrical inspector should check the name plates on all fixed electric space heating equipment to ascertain that the equipment is designed for use on the building's electrical system.

9.1.1 Heating Cable

When the electrical inspector encounters a heating cable installation, he should carefully check to make sure that the cables have been installed in their complete lengths as supplied by the factory. Under no circumstances should the cables be shortened. In the event that the factory labels have been inadvertently removed from the spool on which the cable was shipped from the factory, the cable can be identified by the identification tag on the nonheating leads or by connecting suitable test meters into the circuit.

If joists are run parallel to an outside wall, the cable should cross from one joist space to the next four in. from the wall. A cable should never run nearer than 1¼ in. from the centerline of the joist.

All wiring feeding electric heating cable should, of course, be in accordance with NEC requirements. Furthermore, all branch circuit wiring located above heated areas should be kept above the building insulation. Where it is necessary to enclose the wires in this insulation or between the heated ceiling and a floor above, these wires should be considered as operating in an ambient temperature of 50 °C (112°F). This wiring must have its current-carrying capacity figured according to the correction factors given in the NEC. In no case shall the wiring be supported less than two in. above the heating cable.

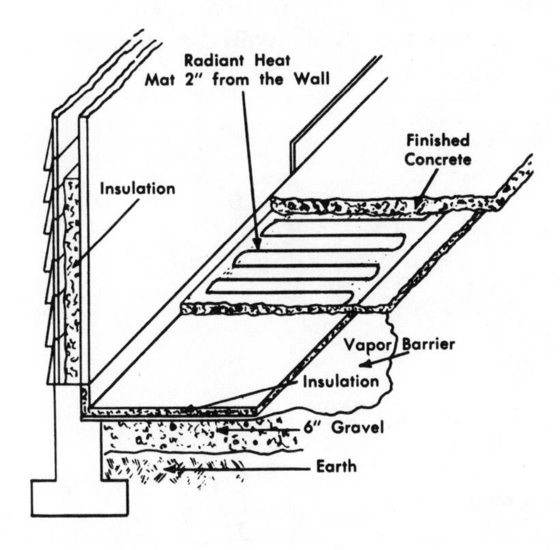

FIG. 9-3. When the electrical inspector enounters a heating cable installation, he should carefully check to make sure the cables have been installed in their complete lengths as supplied by the factory.

9.1.2 Electric Baseboard Heaters

The heating element in electric baseboard heaters should have a maximum operating temperature of 400°F in an ambient temperature of 70°F. A thermal limit control must be supplied and the heaters should be prewired for service connections at either end along with the provision of an approved wire-way. The connection boxes must be of sufficient size for making electrical connections as well as installing required accessories.

It is a good idea to check the manufacturer's installation instructions against the final installation noting any special suggestions or warnings, for example, the minimum distance the units are to be mounted from combustible material, and so forth. All units must be grounded in accordance with Article 250 of the NEC.

9.1.3 Duct Heaters

All duct heaters encountered by the electrical inspector should be UL listed for zero clearance and meet all the applicable requirements of the NEC. Bracket supports for the resistance wire should be reinforced with stiffening ribs and gussets and spaced no more than four in. apart. The heaters should have been tested dielectrically for 1000V plus twice the rated voltage or 2000V, whichever is higher. Heaters rated 150 kW and over should be furnished with heavy duty coils, derated to 35 watts/in. of wire surface to insure long life of the element.

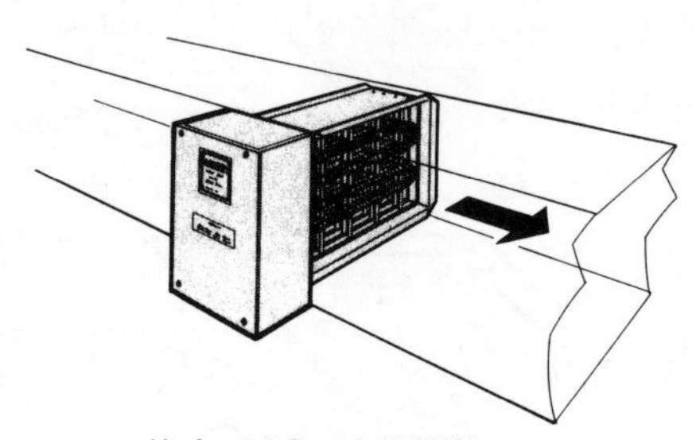

Horizontal Duct Installation

FIG. 9-4. Typical duct heaters.

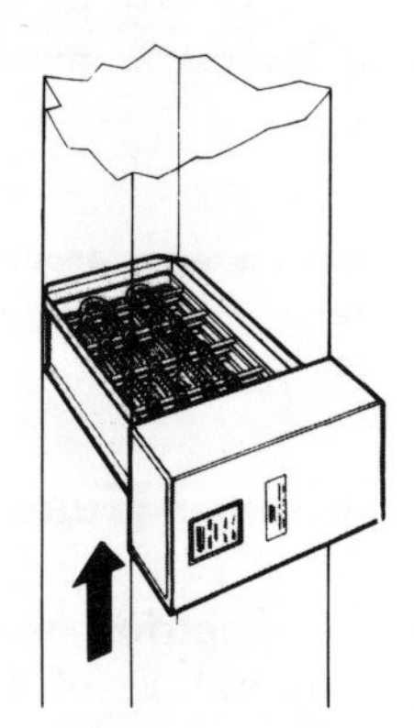

Vertical Duct Installation

All duct heaters should be provided with automatic circuit breakers of the thermal-magnetic or other approved type for overcurrent protection. They must be sized for 80 percent of the circuit and be readily accessible, and also have the following characteristics:

1. One overcurrent device should be provided for each 24-amp circuit.
2. One overcurrent device should be provided for each heating step.
3. One overcurrent device should be provided for the entire heater for those heaters rated 24 amperes or less.

9.2 ELECTRIC ROTATING EQUIPMENT

Electric motors are classified according to size (horsepower); type of application (general purpose, special purpose, etc.); electrical characteristics (voltage and frequency, amperage, ac or dc, type of wiring, phase, etc.); speed, starting, speed control, and torque characteristics; and according to mechanical protection and method of cooling.

Motor controls vary from simple thermal switches to large switchgear-type motor starters. The type of control used for a specific application depends basically upon the type of motor, the type of control circuit, and the location of the controller. Motor controls may be operated manually or automatically by control stations either at the motor or at a point remote from both the controller (starter) and the motor.

When a disconnect means is required for the motor, it may be a separate externally operated safety switch or may be combined in the same enclosure as the starter; in the latter case it is referred to

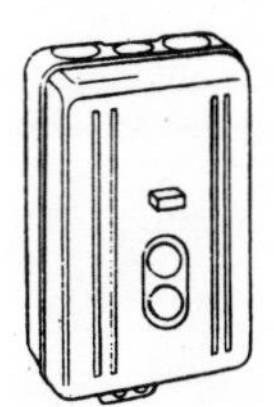
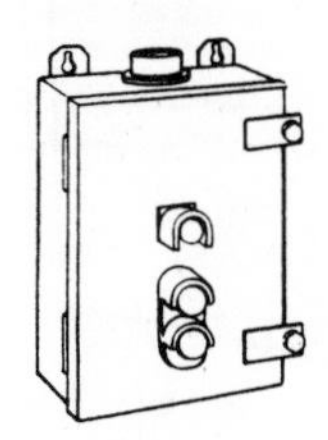
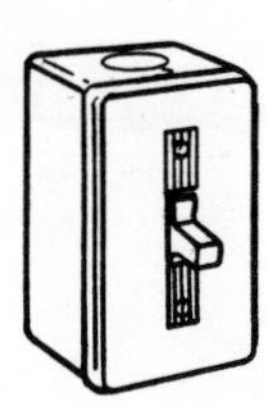

FIG. 9-5. Various types of motor control.

as a combination motor starter. In some instances the overcurrent protection and circuit disconnecting means—for instance, a circuit breaker—may serve as the motor disconnect.

The electrical inspector should do the following things:

1. Check for proper overcurrent protection.

2. Check for proper conductor size and type.

3. Check for proper terminations.

4. Determine if a disconnecting means other than at the branch circuit panel location is necessary, and if so, make sure that it is provided.

9.3 TRANSFORMERS

The main purpose served by transformers is to change the voltage of one section of an electrical system from the voltage of another section of the system, either raising it or lowering it.

A transformer basically consists of two coils of insulated wire wound around an iron core. The coils differ in the number of turns and size of conductors, depending upon the difference in voltage and current capacity of the transformer desired. When a transformer is operating at full capacity, heat is generated and some means of cooling the transformer is necessary. To do so, the transformer can be suspended in a steel tank filled with nonconductive oil, or placed in a metal enclosure provided with ventilation to dissipate the heat.

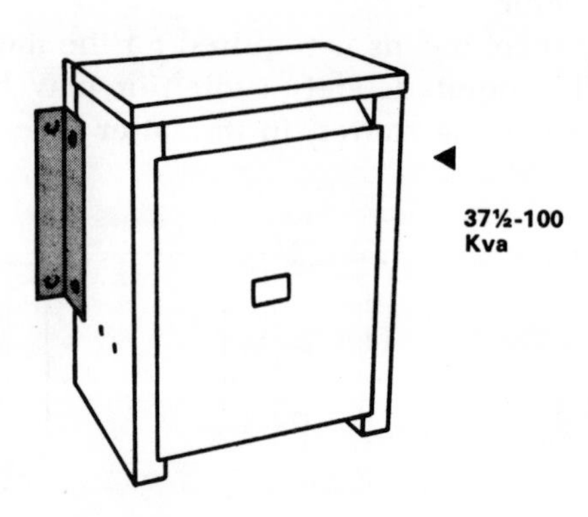

FIG. 9-6. A typical dry-type transformer.

Transformers, together with the related overcurrent protection and disconnecting means, are subject to NEC regulations when installed in or on a building. NEC Article 450 covers most of the requirements for transformers and transformer vaults. The electrical inspector will normally encounter dry type transformers rather than oil cooled transformers and these will be used to reduce 480 volts to 240/120 volts in the majority of installations.

FIG. 9-7. A typical oil filled transformer.

9.4 CAPACITORS, RESISTORS, AND REACTORS

Capacitors are connected to electrical systems to improve the power factor or efficiency of electric power circuits and the system in general. Article 460 of the NEC gives most of the installation regulations.

Resistors and reactors are normally used in conjunction with motor starters and controllers in industrial applications. Article 470 of the NEC gives regulations concerning these items.

All of these items are more often mounted separate from other electrical equipment and are connected to the system or equipment with wiring methods approved by the NEC.

9.5 STORAGE BATTERIES

A storage battery is comprised of one or more rechargeable cells of the lead-acid, nickel-cadmium, or other rechargeable electro-chemical types. Such batteries are often used to provide standby emergency lighting service and to supply power for low current-consuming electric equipment.

FIG. 9-8. Capacitors are connected to electrical systems to improve the power factor or efficiency of the system in general.

FIG. 9-9. Typical storage battery.

When used for standby service, a double-throw manual or automatic switch should be provided. In such areas as hospitals, auditoriums, and other buildings where an electrical failure could prove hazardous to the people within, the switch should be automatic so as to supply a certain amount of lighting should the normal power supply be interrupted.

Wiring and equipment supplied from storage batteries must be subject to the requirements of the NEC applying to wiring and equipment operating at the same voltage. The grounding requirements of Article 250 also apply.

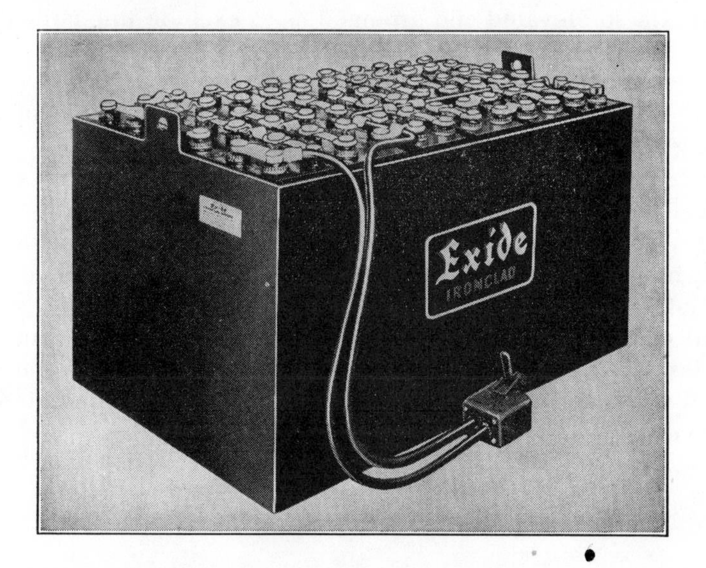

FIG. 9-10. A bank of storage batteries properly located.

The location of storage batteries is very important to the safety of life and property, and the electrical inspector must make certain that all storage batteries are located in areas provided with sufficient diffusion and ventilation of gases from the battery to prevent the accumulation of an explosive mixture. Furthermore, provisions must be made to guard all battery terminals and other live parts as follows:

1. Live parts of batteries and related equipment operating at 50 volts or more must be guarded against accidental contact by approved cabinets or other forms of approved enclosures or by any of the means to follow.

2. Batteries should be located in a room, vault, or similar enclosure that is accessible only to qualified persons.

3. Suitable permanent, substantial partitions or screens should be so arranged that only qualified persons will have access to the space within reach of the live parts. Any opening in such partitions or screens shall be so sized and located that persons are not likely to come into accidental contact with the live parts or to bring conducting objects into contact with them.

4. Batteries should be located on a suitable balcony, gallery, or platform so elevated and arranged as to exclude unqualified persons.

5. Elevation should be 8 ft or more above the floor or other working surface.

In locations where the storage batteries would be exposed to damage, such as where forklifts will be traveling, enclosures or guards must be arranged to prevent any damage. The guards must be of sufficient strength for the conditions.

All entrances to rooms and other guarded locations containing batteries with exposed live parts of 50 volts or more must be marked with conspicuous warning signs forbidding unqualified persons to enter.

Chapter 10
Underground Electrical Systems

Underground electrical distribution systems are being used more frequently each year for power transmission and distribution facilities, illumination facilities, and signal and communication facilities. Any of these systems will fall into two general categories from the standpoint of installation and the electrical inspector's responsibility:

1. Direct burial, where the cables or conductors are buried directly in the earth.

2. Underground raceway systems, in which the conductors are installed at some time after the conduit system is completed.

10.1 DIRECT-BURIAL INSTALLATIONS

Direct-burial installations range from small, single-conductor installations for series airport lighting to large, multiconductor cables for power or communication. In any case, the conductors are installed in the ground either by placing them in an excavated trench which is then backfilled, or by burying them directly by means of a cable plow that opens a furrow, feeds the conductors into the furrow, and closes the furrow over the conductor.

When conductors are buried directly in the earth, the NEC has specific requirements that must be followed, and every electrical inspector should be thoroughly familiar with these requirements or know how to quickly find them in the NEC book. He must furthermore know how to interpret these requirements.

In general, cables of one or more conductors for direct burial in the earth must be of a type approved for the purpose and use. Some types of cables and conductors that may be used for direct-burial wiring include:

- Underground feeder (type UF) cable
- Underground service-entrance cable (Type USE)
- Medium voltage cable (Type MV)
- Mineral-insulated, metal-sheathed cable (Type MI)
- Metal-Clad cable (Type MC)

Of course, there are certain restrictions for all of the types mentioned above, and Chapter 3 of the NEC should always be referred to when there is a question. For example, where it is necessary to prevent damage to the conductors from rocks, slate, or similar materials, or from vehicular traffic, direct buried conductors must be provided with supplementary protection, such as sand, sand and suitable running boards, suitable sleeves, or other approved means.

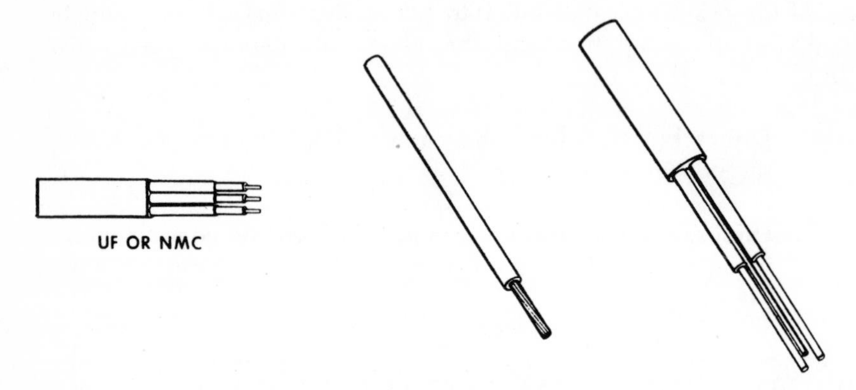

FIG. 10-1. Some types of cables and conductors that may be used for direct-burial wiring installations.

Minimum cover requirements for direct burial cables is 24 in. except for the following conditions:

1. The minimum cover requirements shall be permitted to be reduced by 6 in. for installations where a 2-in. thick concrete pad or equivalent in physical protection is placed in the trench over the underground installation. See the drawing in Fig. 10–2.

2. Residential branch circuits rated 300 volts or less and provided with overcurrent protection of not more than 30 amperes shall be permitted with a cover requirement of 12 in. as shown in Fig. 10–3.

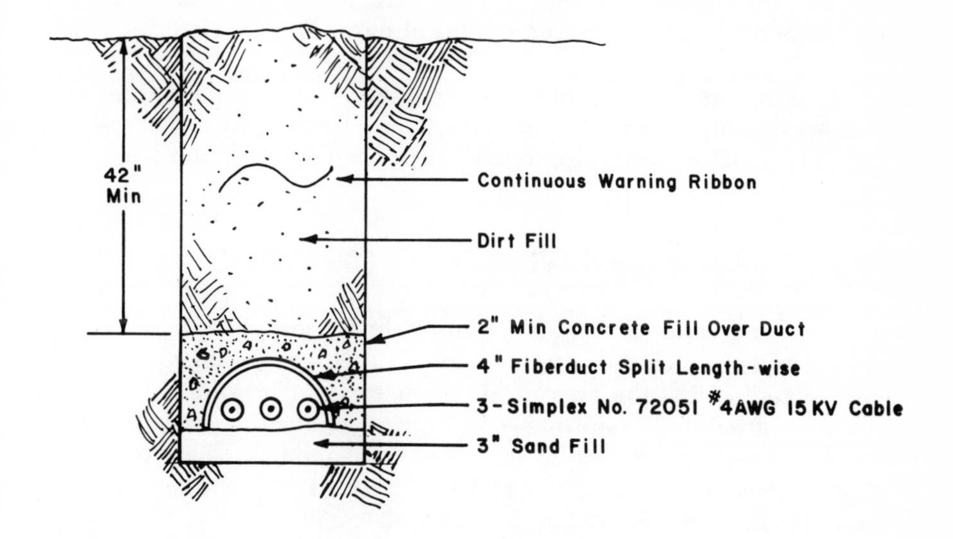

FIG. 10-2. Cross-section of a direct-burial installation.

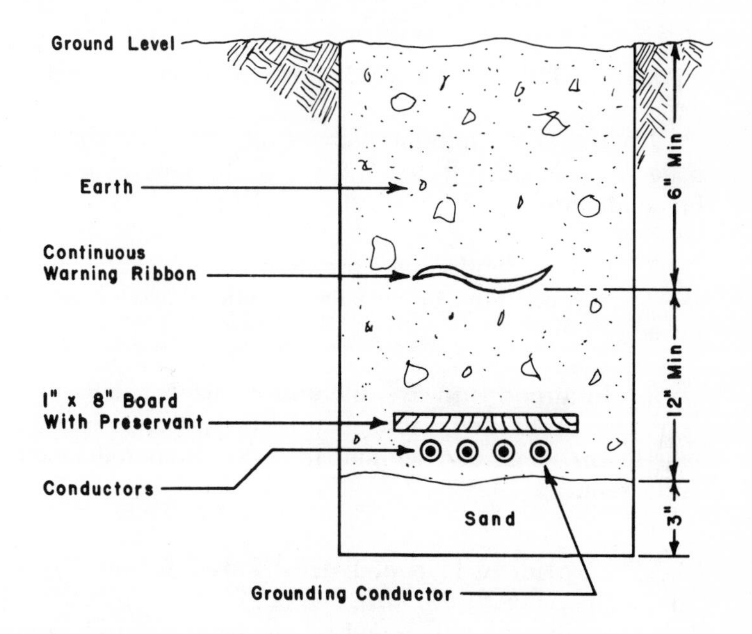

FIG. 10-3. Cross-section of a residential underground wiring installation.

3. Lesser depths are permitted where cables and conductors rise for terminations or splices or where access is otherwise required. See the cable entering the building in Fig. 10-2.

4. Circuits for the control of irrigation and landscape lighting systems that are limited to not more than 30 volts and are installed with Type UF or other approved cable shall be permitted with a minimum cover of 6 in.

10.1.1 Underground Wiring Over 600 Volts

Underground conductors must be suitable for the voltage and conditions under which they are installed. Direct burial cables must be constructed so that the energized conductors are surrounded by effectively grounded, multiple concentric conductors, closely and evenly spaced, or conducting sheath of equivalent ampacity to meet requirement of Section 250-51 of the NEC. Direct buried cables must also meet the following depth requirements:

Circuit Voltage	Depth in Inches
Over 600–22kV	30
Over 22kV–40kV	36
Over 40kV	42

Section 710-3 of the NEC allows the following exceptions:

1. The above minimum cover requirements shall be permitted to be reduced 6 in. for each 2 in. of concrete or equivalent above the conductors.

2. Lesser depths are permitted where cables and conductors rise for terminations or splices or where access is otherwise required.

3. In airport runways, including adjacent defined areas where trespass is prohibited, cable shall be permitted to be buried not less than 18 ins. deep and without raceways, concrete enclosement, or equivalent.

10.1.2 Splicing Direct-Burial Conductors

Underground cables in trenches can be spliced or tapped without the use of splice boxes. However, the splices or taps must be made by methods and with material approved for the purpose.

In most cases, direct burial conductors and cables are spliced or connected at the equipment being fed or in manholes or junction boxes. A direct burial splice box for splicing high-voltage cables (15kV) is shown in Fig. 10-4. On this particular type of splice box, the internal connections do not require insulating with tape or other means. Conductors are connected by soldered-on lugs, bolted to contacts mounted on porcelain supports. This type of splice box is used where it is desirable or required to splice continuing runs of underground cable at points other than in junction boxes, manholes, or at point of contact with electrical equipment. The electrical inspector should require that markers be installed to indicate the location of buried cable splices.

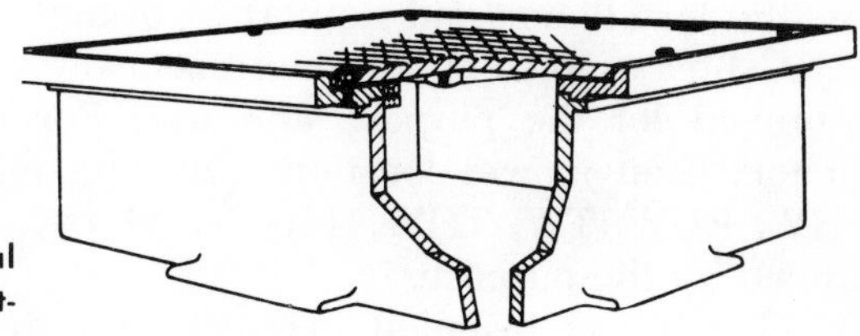

FIG. 10-4. A direct-burial splice box for splicing high-voltage cables.

10.2 UNDERGROUND RACEWAY SYSTEMS

Underground raceway systems consist of manholes or junction boxes and connecting runs of one or more rigid, PVC, or fiber ducts placed in trenches and usually encased with a concrete envelope. The size and number of raceways depend upon the use of the system, number of conductors, and spare raceway capacity desired. The separation of the raceways depends upon the mechanical strength of the system and the voltage that it will carry.

The conduits provide routing for the conductors, while the manholes and junction boxes are provided to allow for splicing of continuous conductor runs, and to facilitate the installation of continuous conductor runs, the junction of lateral runs, and the installation of operating equipment.

The size, shape, and type of manholes are usually governed by the number of conductors to be run into them. Some will be small and relatively shallow, and these are often referred to as junction boxes or handholes.

In general, manholes consist of a main chamber, an access throat from the top of the main chamber to the grade level above, a cast-iron ring and cover, possibly some provision for drainage, race-

FIG. 10-5. Cross-section of an underground raceway system.

way entrances, access ladder, and miscellaneous hardware to facilitate the installation and supporting of the conductors. See Fig. 10–6.

Cables used in underground raceway systems must be of a type approved for the purpose and use. For example, insulated conductors used in wet locations must be either lead-covered, types RHW, RUW, TW, THW, THWN, XHHN, or of a similar type approved for the purpose.

Conductors installed in raceway systems also have minimum cover requirements. The NEC defines "cover" as the distance between the top surface of the conduit system and the finished grade. The minimum requirements for systems under 600 volts are as follows:

Wiring Method	Minimum Burial (inches)
Rigid metal conduit	6
Intermediate metal conduit	6
Rigid nonmetallic conduit (PVC, etc.)	18
Other approved raceways	18

Exceptions to the above minimums are as follows:

1. The minimum cover requirements shall not apply to conduits or other raceways located under a building or exterior concrete slab not less than 4 in. in thickness and extending not less than 6 in. beyond the underground installation.

2. Areas subject to heavy vehicular traffic, such as thoroughfares, shall have a minimum cover of 24 in.

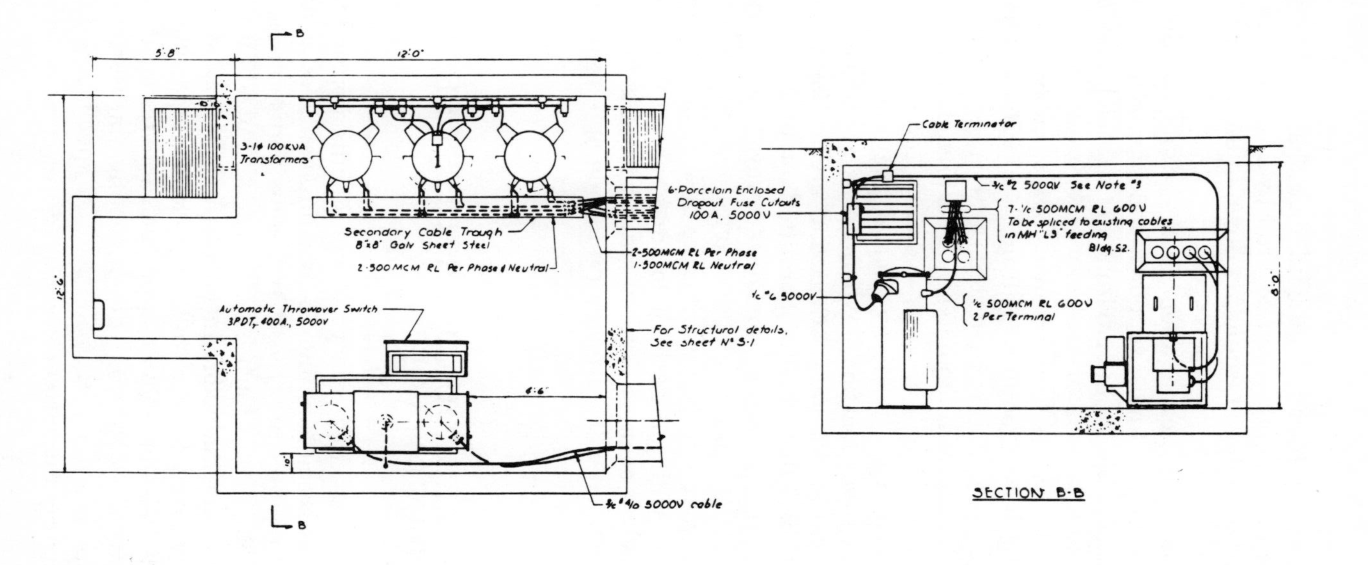

FIG. 10-6. Plan and elevation view of a typical manhole.

3. Lesser depths are permitted where cables and conductors rise for terminations or splices or where access is otherwise required.

4. Raceways installed in solid rock shall be permitted to be buried at a lesser depth when covered by 2 in. or more of concrete over the installation and extending down to the rock surface.

Metal raceways, cable armor, boxes, cable sheathing, cabinets, elbows, couplings, fittings, supports, and support hardware must be of materials suitable for the environment in which they are to be installed.

Ferrous raceways, cable armor, boxes, cable sheathing, cabinets, metal elbows, couplings, fittings, supports, and support hardware must be suitably protected against corrosion inside and outside (except threads at joints) by a coating of approved corrosion-resistant material such as zinc, cadmium, or enamel. Where protected from corrosion solely by enamel, they must not be used out of doors or in wet locations. When boxes or cabinets have an approved system or organic coatings and are marked "Raintight," or "Outdoor Type," they shall be permitted out of doors.

Ferrous or nonferrous metal raceways, cable armor, boxes, cable sheathing, cabinets, elbows, couplings, fittings, supports, and support hardware will be permitted to be installed in concrete or in direct contact with the earth, or in areas subject to severe corrosive influences when made of material judged suitable for the condition, or when provided with corrosion protection approved for the condition.

10.3 UNDERGROUND CIRCUITS OPERATING AT MORE THAN 600 VOLTS

On many projects, high voltage power circuits are used to feed such equipment as gaseous tubing, X-ray equipment, and certain types of motors and equipment, as well as feeders and services. Conductors used for such systems must be of a type approved for the operating voltage, and the conductors installed in a raceway system or cable must be approved for the voltage. Lead-covered, high-voltage cables must be terminated with approved cable terminating devices.

Overcurrent protection, disconnecting, measuring, and operating equipment must be manufactured specifically for the operating voltages of the system.

10.3.1 Splicing of High-Voltage Cables

When required, high-voltage cables are spliced in manholes or underground cable junction boxes. Cables, however, are not always spliced in every manhole; in some instances cable may be pulled through one or more manholes with sufficient slack to allow racking.

The splicing of cables, particularly lead covered cables, is a specialized operation that should be performed only by capable workers who have acquired the knowledge of and the ability to perform such operations. In general, these splices must be made in such a manner as to properly insulate the conductors for the system voltage and to prevent the entrance of moisture into the cable insulation or into the equipment. Furthermore, correct splices provide a permanently secure electrical and mechanical joining of the conductors or connection to the equipment terminals. Sometimes, project specifications require that the exposed sections of cable in manholes be fire-proofed with a wrapping of asbestos tape impregnated with an asbestos cement.

When the electrical inspector encounters projects dealing with the splicing of high-voltage conductors, he should make certain that the person or persons performing the splicing are qualified. Even then, once the splice is completed, tests should be made to insure against leaks in the insulation before the system is energized.

High-voltage cable splicers are normally registered with the I.B.E.W., if members; open-shop splicers are usually registered with the local inspector's office.

During underground electrical construction work, the electrical inspector should be present—even for a short time—when the following operations are taking place:

1. During the construction of side wall forms.

2. When concrete mats are provided for the first tier of raceway.

3. After the tying or banding fiberducts together and before the concrete is poured.

4. Backfilling and replacing the paved surface, when required, over the raceway installation.

5. Pulling the conductors.

6. Splicing the conductors.

Chapter 11
Overhead Electrical Systems

Outside overhead electrical systems include such work as wooden pole overhead line construction, steel tower overhead line construction, substation and switchyard construction, and overhead trolley systems. Most NEC requirements for the construction of overhead systems are found in Chapter 2 namely, Article 225—Outside Branch Circuits and Feeders. Application of other NEC articles, including additional requirements to specific cases of equipment and conductors, are as follows:

Branch circuits	Article 210
Remote control and power-limited circuits	Article 725
Communication circuits	Article 800
Community antenna television	Article 820
Conductors	Article 310
Electric irrigation machines	Article 675
Electric signs	Article 600
Feeders	Article 215
Fire protective systems	Article 760
Fixed snow-melting equipment	Article 426
Fixtures	Article 410
Grounding	Article 250

Hazardous locations	Article 500
Marinas and boatyards	Article 510
Over 600 volts, general	Article 710
Overcurrent protection	Article 240
Radio and television equipment	Article 810

11.1 OUTSIDE CIRCUITS AND CONDUCTORS

The load on outdoor branch circuits and feeders is determined as follows:

1. The continuous load supplied by a branch circuit must not exceed 80 percent of the branch-circuit rating except where branch-circuit conductors have been derated in accordance with Note 8 to Tables 310–16 through 310–19 of the NEC, or where the assembly—including overcurrent devices—is listed for continuous operation of 100 percent of its rating.

2. A unit load of not less than that specified in Table 220–2(b) (Chapter 7, Section 7.1 of this book) for occupancies listed therein shall constitute the minimum lighting load for each sq ft of floor area of a building.

3. In all occupancies the minimum load for each outlet for general-use receptacles and outlets not used for general illumination must be not less than the following, the loads shown being based on nominal branch-circuit voltages.

   (a) Outlet for a specific appliance or other load except for a motor load must be at the ampere rating of appliance or load served.

   (b) Outlets for motor loads must be calculated according to Sections 430–22, 430–24, and Article 400 of the NEC.

   (c) Outlet for heavy-duty lamphold must be rated at 600 va (watts).

   (d) Other outlets calculated at 180 va (watts) per outlet. A conventional duplex receptacle would then be calculated at 360 watts per outlet since there are two receptacles per outlet.

In most cases, overhead conductors may be run bare except that within 10 ft of any building or other structure, open wiring on insulators must be insulated or covered. Conductors in cables or raceways, except Type MI cable, must be of the rubber-covered type or thermoplastic type and in wet locations, conductors must comply with Section 310–7 of the NEC. Conductors for festoon lighting must also be of the rubber-covered or thermoplastic type. (Festoon lighting is a string of outdoor lights suspended between two points more than 15 ft apart.)

The size of overhead conductors for outside wiring is based on the load carried by them. Size conductors must be in accordance with Tables 310–16 through 310–19 of the NEC. In no case, however, should overhead conductors be smaller than No. 10 AWG for spans up to 50 ft in length and No. 8 AWG for longer spans. For voltages over 600 volts, No. 6 AWG is the smallest size conductor permitted where open individual conductors are used, or No. 8 AWG when the conductors are in an approved cable assembly.

Except where supported by messenger wires, overhead conductors for festoon lighting must not be smaller than No. 12.

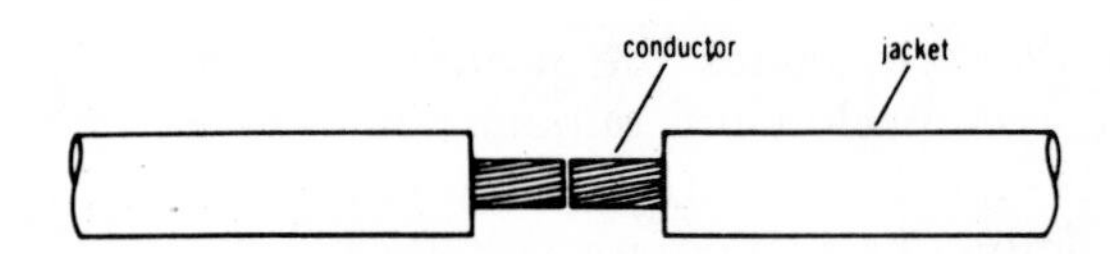

FIG. 11-1. Conductors for festoon lighting must be of the rubber-covered or thermal plastic type.

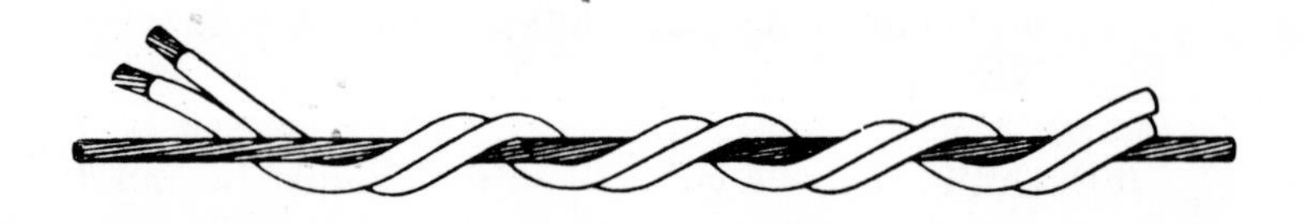

FIG. 11-2. Overhead conductors supported by messenger wires.

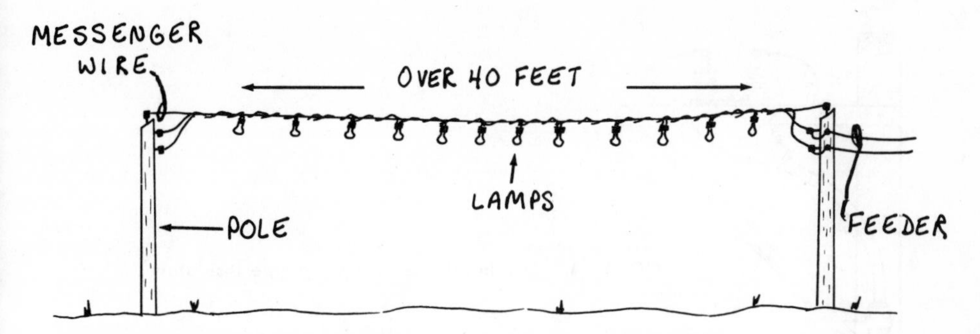

FIG. 11-3. Where festoon lighting spans exceed 40 ft in length, the conductors must be supported by messenger wire.

11.2 OUTDOOR LIGHTING EQUIPMENT

Overhead branch circuits supplying lampholder or lighting fixtures mounted on the outside of buildings or on poles or structures for area illumination of residential, commercial, or industrial property must not exceed 150 volts to ground except as follows:

> *When lighting fixtures are mounted on the outside of buildings classified as industrial establishments, office buildings, schools, stores, and other commercial or public buildings, the voltage to ground may be increased to 300 volts provided the fixtures are not less than 8 ft above grade (or other surface accessible to individuals) and the fixtures are not less than 3 ft from windows, platforms, fire escapes, and the like.*

The disconnecting means and overcurrent protection for branch circuits supplying outdoor lighting fixtures are the same as for conventional branch circuits.

For festoon lighting where the spans exceed 40 ft in length, the conductors must be supported by messenger wire as shown in Fig. 11-3. Note that the messenger wire is supported by strain insulators. Festoon lighting conductors or messenger wires must not be attached to any fire escape, downspout, or plumbing equipment.

11.3 CONDUCTORS ON POLES

Overhead lines using pin-type insulators, insulated clevices, light-weight groups of suspension insulators, or insulator racks are usu-

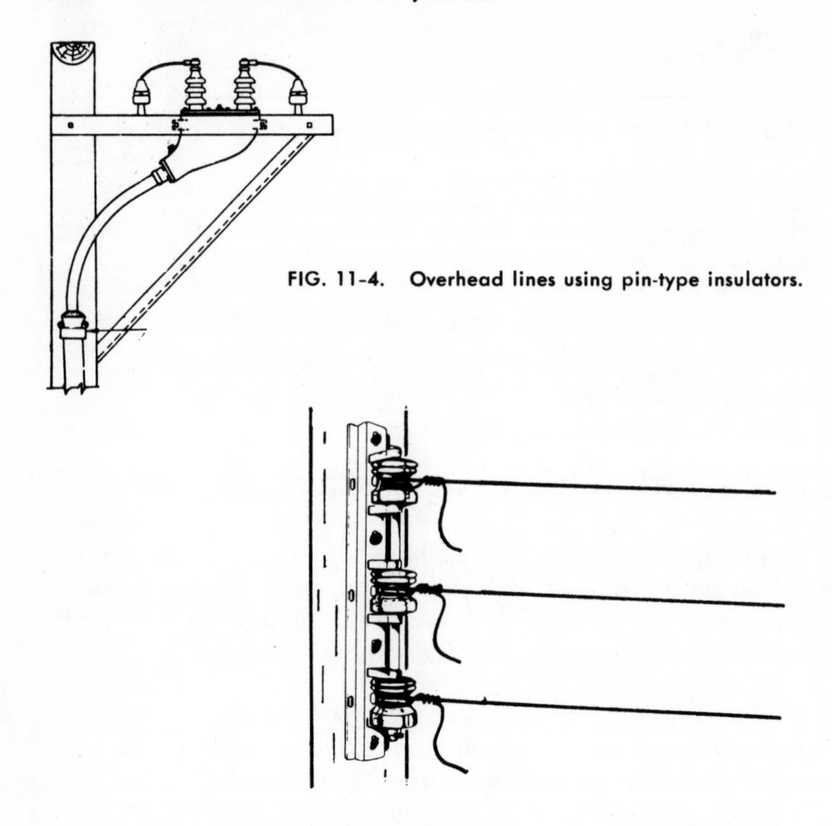

FIG. 11-4. Overhead lines using pin-type insulators.

FIG. 11-5. Overhead lines using insulator racks to support the conductors.

ally constructed with individual poles. The basic components of wooden pole construction consist of the poles and related pole holes and auxiliary supports, cross arms, cross-arm braces, multipole structure cross braces, insulators, and related hardware as shown in Fig. 11-6. Other equipment includes transformers and switching equipment, equipment platforms, and pole top structures.

The height or length of poles depends primarily upon their use and the terrain on which they are erected. Such items as voltage of the system, number of circuits, type and extent of equipment to be mounted on the poles, changes in direction of the line, length of conductor span, and so on, are all important factors in determining the height of wooden poles for electrical distribution.

The usual lengths of wooden poles range from 20 to 60 ft on most installations. The shorter ones will be used on single-circuit low-voltage distribution lines and communication systems such as electric lines between buildings on a property. Longer poles may be required for transmission lines and certain applications of distribution lines. Exceptionally long poles may be required for power-line river or canyon crossings and certain recreation floodlighting applications such as football and baseball stadium lighting, and so on.

The electrical inspector should check the class of poles installed against the designer's specifications as as well as the type of preservative treatment, minimum length, foundations, and the like.

Conductors installed on poles must have a separation of not less than one ft where the conductors are not installed on racks or brackets. Conductors supported on poles must also be placed to provide a horizontal climbing space not less than the following:

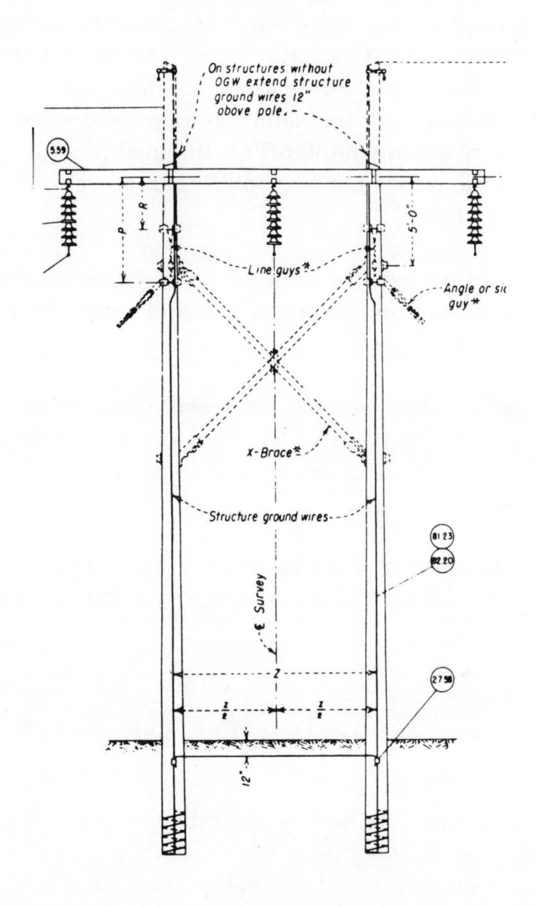

FIG. 11-6. The basic components of wooden pole construction.

1. Power conductors, below communication conductors	30 in.
2. Power conductors alone or above communication conductors:	
300 volts or less	24 in.
Over 300 volts	30 in.
3. Communication conductors below power conductors	30 in.
4. Communication conductors alone	no requirement

Open conductors must be supported on glass or porcelain knobs, racks, brackets, or strain insulators. When overhead conductors pass over a roof, they must be securely supported by substantial structures and if possible, they should be independent of the building. Where the conductors are attached to the building, they must be installed in the same manner as service entrance conductors as described in Chapter 5 of this book.

Open conductors of not over 600 volts must conform to the following requirements:

1. Conductors must be at least 10 ft above finished grade and sidewalks, or from any platform or projection from which they might be reached.

2. Where conductors pass over residential driveways and commercial areas such as parking lots and drive-in establishments not subject to truck traffic, the lines must be a minimum of 12 ft above finished grade.

3. The conductors must be at least 15 ft above finished grade over commercial areas, parking lots, agricultural, or other areas subject to truck traffic.

4. Conductors must be at least 18 ft above grade over public streets, alleys, roads, and driveways on other than residential property.

5. Conductors must have a clearance of not less than 8 ft from the highest point of roofs over which they pass except when they meet the requirements listed in Section 225–19(a) of the NEC.

6. Conductors not attached to a building must have a minimum horizontal clearance of 3 ft.

7. Conductors attached to buildings must be kept at least 3 ft from windows, doors, porches, fire escapes, or similar locations. Conductors run above the top level of a window are considered to be out of reach from that window.

FIG. 11-7. Conductors run above the top level of a window are considered to be out of reach from that window.

11.4 OVERHEAD COMMUNICATION SYSTEMS

Interior communication systems were described in Chapter 8 (Communication Equipment). This section is concerned with central-station type systems—those normally installed overhead on system poles or jointly with power distribution systems.

Protective devices are required in communication systems entering buildings to prevent the higher voltage power current from passing into the building in case of accidental contact with power systems or lightning strikes. When installed in raceways, communication and signal circuits must be installed in a separate raceway from power circuits; both must be grounded in accordance with Article 250 of the NEC. Communication and power conductors must not enter the same outlet boxes unless the conductors are separated by a partition or unless the power circuit conductors are introduced soley for power supply to communication equipment or for connection to remote-control equipment.

When communication systems are installed overhead, proper clearances from power circuits must be maintained as well as over streets, sidewalks, driveways, and so forth, as described in Section 11.3.

Communication system conductors vary in size, type of insulation, and combination of conductors, depending upon the system requirements. They are installed as single conductors, as two or more conductors twisted together, or in a cable assembly, often containing a large number of small-sized identifiable conductors.

Conductors are connected directly to the equipment or to terminal blocks. Lead-sheathed cables are spliced when necessary to make long continuous runs, and are usually terminated in special terminal boxes to prevent moisture from entering the insulation. Lead-sheathed cables are usually installed in overhead systems suspended from steel messenger cables.

Other points covering the installation of communicating circuits are contained in Chapter 8 of the NEC.

11.5 RADIO AND TELEVISION EQUIPMENT

Electrical inspectors are often called upon to inspect radio and television transmitting and receiving equipment as well as outside antenna systems and structures supporting antenna systems. Aside from the technical design and installation of the system, the electrical inspector should be familiar with the problems of mechanical strength of the conductors and supporting structures and their installation. In general, antenna systems should be constructed so as to avoid the hazard of contact with power lines and hazard to life and property in the case of mechanical failure of the structure or conductors.

Adequate protection from lightning and proper grounding, both for safety and satisfactory system operation, is required.

The electrical inspector must be familiar with all local codes and ordinances, as these vary considerably from one location to another in their requirements for the installation of antenna systems for both commercial transmitting and local reception. Specific NEC requirements are found in Article 810 of the NEC.

Chapter 12
Miscellaneous Electrical Systems

The majority of electrical inspectors will most often encounter the electrical systems described in previous chapters. However, from time to time the inspector will be called to inspect relatively unusual situations like those described in this chapter. When such projects are encountered, the electrical inspector should review the Articles pertaining to the situation in the NEC book, and other reference books or authorities if the NEC requirements are vague.

During the time of inspection, if certain hazy areas are encountered, the inspector should make notes and perhaps take photos of the situation for further research before either approving or turning down a project. The following data will give the electrical inspector guidelines concerning some of those situations so that he will be able to approach his work more intelligently.

12.1 REMOTE CONTROL POWER AND SIGNAL CIRCUITS

The electrical inspector will sometimes encounter power and lighting equipment controlled at a location remote from the equipment. In such cases, the electrical circuits are controlled—turned off and on—by means of relays or contactors. Most relays are energized (closed or opened) by the relay coil or by electronic components. A relatively small amount of current flows through the relay control circuit, whereas a large amount of current may flow through the circuit operating the equipment. Therefore, the relay control circuit conductors may be smaller than the minimum requirement for the lighting and power circuits.

The relay control circuit may obtain its current from connection directly to the line side of the relay, in which case the amount of power flowing in the circuit is not limited, or through a current-limiting device, such as a current-limiting transformer, or an overcurrent protective device.

The installation requirements of such control circuits are fully covered in Article 725 of the NEC. The various circuits are segregated into Class 1, 2, and 3 circuits.

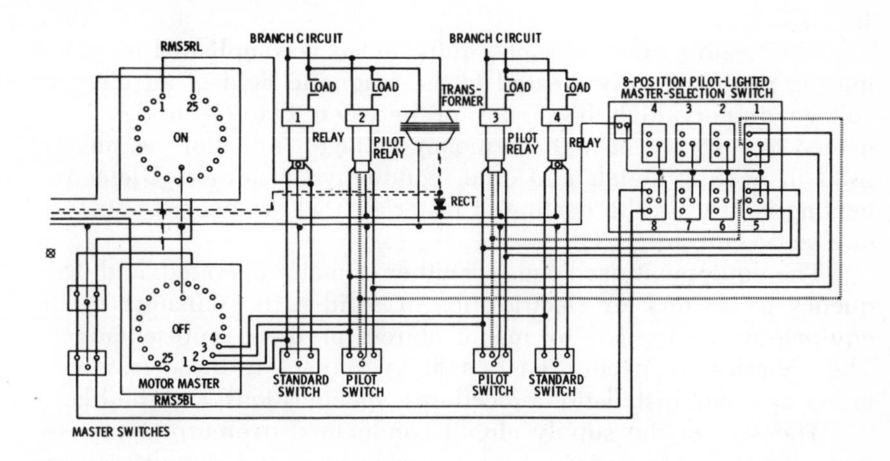

FIG. 12-1. Low-voltage, remote-control system used to control lighting circuits.

12.2 MACHINE TOOLS

Control wiring and feeder connection terminals on nonportable, electrically driven machines are usually installed at the factory. In most cases, due to the areas in which the equipment is used, the wiring method is restricted to rigid conduit except for short lengths of flexible conduit where necessary for final connection to the equipment. Continuously moving parts of the machine are interconnected with approved type, extra flexible, nonmetallic covered cable. The size of the conductors, type and mounting of control equipment, overcurrent protection, and groundings are covered in Article 670 of the NEC.

The electric supply for metalworking machines may be from conventional branch circuits or feeders or in the form of bus ducts. This latter method provides a very flexible type of installation allowing the moving of machines from one part of the plant to another and their reconnection to another part of the the bus duct system, eliminating changes of raceway wiring. The installation of busways is covered in Article 364 of the NEC.

12.3 INDUCTION AND DIELECTRIC HEATING EQUIPMENT

The wiring for and connection of induction and dielectric heat generating equipment used in industrial and scientific applications (but not for medical or dental applications) are covered in Article 665 of the NEC.

The heating effect of such equipment is accomplished by placing the materials to be heated in the magnetic field of an electric voltage of very high frequency or between two electrodes connected to a source of high frequency voltage. Induction heating is used in heating metals and other conductive materials. Dielectric heating is used in the heating of materials that are poor conductors of electric current.

The equipment used consists either of motor-operated, high-frequency generators, or electric tube or solid-state oscillators. Such equipment is supplied by manufacturers or their representatives. The electrical inspector can benefit by contacting these manufacturers to obtain installation procedures, specifications, and the like.

The size of the supply circuit conductors, overcurrent protection, disconnecting means, type of grounding, and output circuits are covered in Article 665 of the NEC.

12.4 ELECTRIC WELDERS

Electric welding equipment is normally treated as a piece of industrial power equipment for which are provided branch circuits adequate for the current and voltage of the equipment. Certain specific conditions, however, apply to circuits feeding AC transformer and dc rectifier arc welders, motor-generator arc welders, resistance welders, and so forth; the requirements are found in Article 630 of the NEC.

The electrical inspector should be aware that electrical arc welders require a relatively large amount of current at a relatively low voltage delivered to the welding electrode. The welding is accomplished with the melting and fusing of the welding rod or electrode to the material being welded. The reduction of the voltage and the increase in current are accomplished through the use of transformers or motor generators.

Resistance welding is accomplished by the melting and fusing of the metals due to the passage of a large amount of current from one electrode to another through the material being welded.

In arc welding, the current flows continuously as long as the welding rod is in contact with the metal being welded. In resistance welding, the flow of the current is interrupted intermittently either manually or automatically, the length of time depending upon the type of weld being made.

FIG. 12-2. The electrical inspector should refer to Article 630 of the NEC when welding equipment is encountered.

12.5 CRANES AND HOISTS

Crane and hoist equipment is usually furnished and mechanically installed by crane manufacturing companies or their representatives. When the electrical inspector is required to inspect such an installation, he should become thoroughly familiar with Article 610 of the NEC.

Crane or hoist wiring consists of the control and operating circuits on the equipment itself and the contact conductors or flexible conductors supplying electric power to the equipment. Electricians on the job normally install the feeder circuit at a point of connection to the contact conductors or flexible cables, including overcurrent protection and disconnecting means.

The motor control equipment, control and operating circuits, and the bridge contact conductors are usually furnished and installed by the manufacturer; sometimes, it is the responsibility of the owner or electrical contractor. The electrical inspector should determine to what extent he is responsible for the installation and act accordingly.

12.6 SIGNS AND OUTLINE LIGHTING

Electric signs and outline lighting are covered in Article 600 of the NEC. Such equipment is used for decorative and advertising purposes and is usually self-contained and attached to, rather than a part of, the building or wiring system. The inspector must make sure that overcurrent protection is provided for the circuit feeding the piece of equipment and also that an approved disconnecting means is provided. All such equipment must also be grounded in accordance with Article 250 of the NEC.

FIG. 12-3. Rules governing the installation of sign lighting may be found in Article 600 of the NEC.

Gaseous (neon) tubing systems require the use of step-up transformers to provide the necessary high operating voltages. Therefore, the secondary amperage or current is usually relatively small. However, because of the high voltage, secondary conductors must have an insulation adequate for operating properly and all terminals must be protected from or inaccessible to unqualified persons. In most instances, the sign equipment is furnished and installed by sign manufacturing companies and the electrical contractor furnishes and supplies the feeder circuits.

Such lighting equipment is illuminated by incandescent, or electric-discharge lamps. Controllers known as flashers are often used to obtain a blinking effect for further decoration or to gain attention.

12.7 THEATERS AND SIMILAR LOCATIONS

The NEC recognizes that hazards to life and property due to fire and panic exist in theaters and similar types of buildings and has incorporated certain requirements in addition to those of the usual commercial wiring installations. These requirements include:

1. Proper wiring of motion picture projection rooms (Art. 540).

2. Heat-resistant insulated conductors for certain lighting equipment (Art. 520–43).

3. Adequate guarding and protection of the stage panelboards, and proper control and overcurrent protection of all circuits (Art. 520–22).

4. Proper type and wiring of lighting dimmers (Art. 520–25 and 520–53).

5. Use of proper types of receptacles and flexible cables for stage lighting equipment (Art. 520–45).

6. Proper stage flue damper control (Art. 520–49).

7. Proper dressing room wiring and control (Art. 520–71, 72, and 73).

8. Fireproof projection rooms with automatic projector port closures, ventilating equipment, emergency lighting, guarded work lights, and proper location of related equipment (Art. 540).

Outdoor and drive-in type motion picture theaters do not present the inherent hazards of enclosed auditoriums. However, the projection rooms must be properly ventilated and wired for the protection of the operating personnel.

12.8 MOBILE HOME AND TRAILER PARK INSTALLATIONS

In recent years, mobile home and travel trailer parks have increased in numbers to the point where the NEC found it necessary to add additional requirements to the Code book. Therefore, in the 1971 edition of the NEC, Article 550—Mobile Homes and Mobile Home

Parks and Article 551—Recreation Vehicles and Parks were added. Subsequent editions have retained and supplemented these Articles.

In general, the pertinent provisions of these Articles cover the electrical conductors and equipment installed within or on mobile homes and recreation vehicles, as well as the means of connecting the units to a supply of electricity.

12.8.1 Sizing Electrical Services for Mobile Homes

The electric service for a mobile home may be installed either underground or overhead, but the point of attachment must be a pole or poser pedestal located adjacent to the mobile home and *not* mounted on or in the mobile home. The power supply to the mobile home itself is then provided by a feeder assembly consisting of not more than 3 mobile home power cords, each rated for at least 50 amperes for a permanently installed circuit.

The NEC gives specific instructions for determining the size of the supply-cord and the distribution-panel load for each feeder assembly for each mobile home. The calculations are based on the size of the mobile home as well as the small appliance circuits and other electrical equipment that will be connected to the service.

Lighting loads are computed on the basis of the mobile home's area: width times length (outside dimensions exclusive of coupler) times 3 watts per ft^2; that is,

$$\text{Length} \times \text{width} \times 3 = \underline{\hspace{3cm}} \text{ lighting watts}$$

Small appliance loads are computed on the basis of the number of circuits times 1500 watts for each 20-ampere appliance receptacle circuit; that is,

$$\text{Number of circuits} \times 1500 = \underline{\hspace{3cm}} \text{ small appliance watts}$$

Once these two loads are determined, the sum of the two gives the total load in watts. However, there is a *diversity* (demand) factor which may be applied to this total in sizing the service and power cord. The first 3,000 watts (obtained from the previous calculation) is rated at 100 percent; the remaining watts should be multiplied by 0.35 (35 percent). The total wattage so obtained is divided by the feeder voltage to obtain the service size in amperes.

If other electrical loads are to be used in the mobile home, the nameplate rating of each must also be determined and entered in the summation. Therefore, if the electrical inspector needs to determine the total load for a mobile home power supply to check the installation, he should calculate:

1. Lighting and small appliance load as discussed previously.

2. Nameplate amperes for motors and heater loads, including exhaust fans, air conditioners, electric heaters, etc. Obviously, air conditioners and heaters will not operate simultaneously, so only the larger of the two needs to be included in the total load figures. Also multiply the largest motor nameplate rating by 1.25 and add the answer in the calculations.

3. Total of nameplate amperes for any garbage disposals, dishwashers, electric water heaters, clothes dryers, cooking units, etc. Where the number of these appliances exceeds 3, use 75 percent of the total load.

4. Derive amperes for free-standing ranges (as distinguished from separate ovens and cooking units) by dividing values shown in the following table by the voltage between phases.

TABLE 12–1
POWER DEMAND FACTORS
FOR FREE-STANDING ELECTRIC RANGE

Nameplate Rating	Use
10,000 watts or less	80 percent of rating
12,001–12,500	8,000 watts
12,501–13,500	8,400 watts
13,501–14,500	8,800 watts
14,501–15,500	9,200 watts
15,501–16,500	9,600 watts
16,501–17,500	10,000 watts

5. If outlets or circuits are provided for other than factory-installed appliances, include the anticipated load.

12.8.2 Sizing Electric Services and Feeders for Parks

A minimum of 75 percent of all recreation vehicle park lots with electrical service equipment must be equipped with both a 20-ampere, 125-volt receptacle, and a 30-ampere, 125-volt receptacle. The remainder of all lots with electrical service equipment may be equipped with only a 20-ampere, 125-volt receptacle.

Since most travel trailers and recreation vehicles of recent manufacture are equipped with 30-ampere receptacles, an acceptable arrangement is to install a power pedestal (where possible) in the corner of four lots so that four different vehicles can use the same pedestal. Such an arrangement would require three 30-ampere receptacles and one 20-ampere receptacle to comply with Section 551–44 of the NEC. A wiring diagram showing the distribution system of a park electrical system serving 20 recreation vehicle lots is shown in Fig. 12–4.

Electric service and feeders must be calculated on the basis of not less than 3,600 watts per lot equipped with both 20-ampere and 30-ampere supply facilities and 2,400 watts per lot equipped with only 20-ampere supply facilities. The demand factors set forth in Table 551–44 of the NEC are the minimum allowable demand factors that may be used in calculating load for service and feeders.

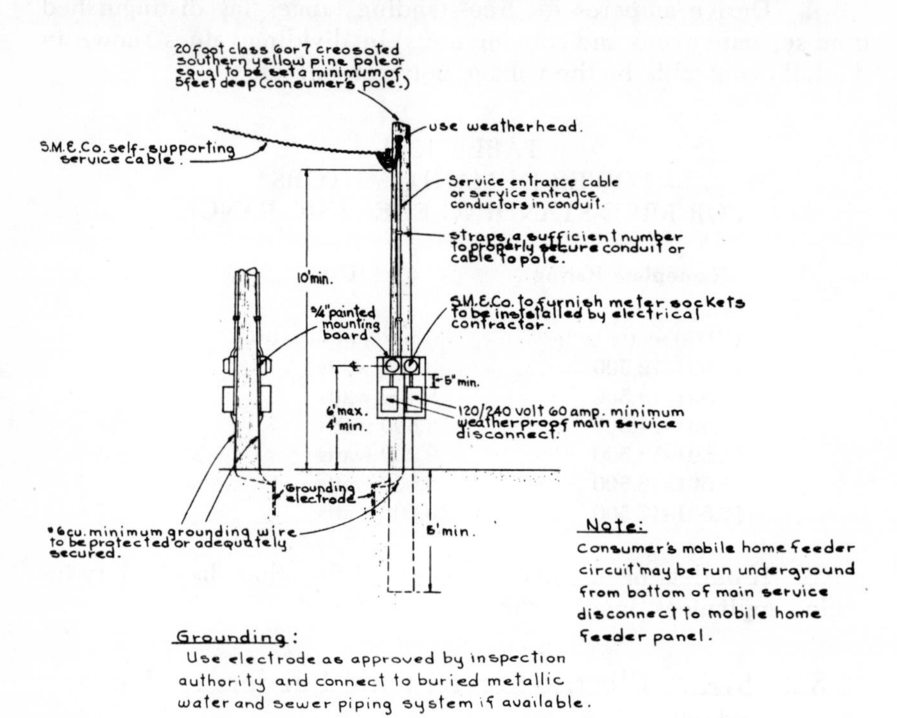

FIG. 12-4. A wiring diagram showing the distribution system for a travel trailer park.

Chapter 13
Scheduling

Scheduling is important to all electrical inspectors regardless of the capacity in which they are employed. Inspectors working for architects and consulting engineers must plan their schedules to insure that they will be on the job site during each sequence of construction so as not to miss any details called for in the plans and specifications. Furthermore, they should not hold up construction or progress payments.

Inspectors employed by state or city municipal agencies should plan their scheduling so as not to hold up construction progress. For example, a day or two delay inspecting the rough wiring in a deck can not only delay the electrical contractor, but can also hold up the general contractor from pouring the concrete—involving the lost time of perhaps several dozen men.

13.1 THE BASIS OF SCHEDULING

In general, the basis of scheduling the progress of the electrical work on a given project is simply the anticipation of what will be required of the electrical contractor in the future and when it will be accomplished.

Most electrical contractors start planning their own schedules prior to starting actual installations on the job; therefore the electrical inspector can save much time if the electrical contractor is consulted frequently.

For the smaller projects, the scheduling may consist only of a mental analysis of the project in question; that is, the inspector should find out the time and date the project will be started and approximately when the various inspection phases will occur. The inspector can then note on his calendar these dates in order to plan his own time in order not to hold up construction.

The larger projects will normally require considerably more formal planning and scheduling for both the contractors and inspectors. Degrees of planning and scheduling can range from notations on the drawings to the use of computers for a more orderly, convenient, and faster procedure. The format can also vary from print-outs in tabulations to various types of graphical charts or a combination of the two.

13.1.1 The Bar Graph

The traditional method of demonstrating an intended schedule has been the bar graph like the one shown in Fig. 13–1. In this format, the work items are described along the vertical axis, while the length of time to accomplish the work is shown on the horizontal axis, to a time scale. For the schedule to be truly effective, it must show enough detail along the vertical axis to reflect exact working conditions. For example, "rough-in deck outlets" is an insufficient description to use on, say, a twelve-story building. However, when the data are broken down into finer detail, a more effective means is provided for both construction and inspections.

If the above example were broken down in the following way, the electrical inspector could quite accurately budget his time to be on the job site at the required time so as not to hold up construction or to waste his own time waiting for the work to be completed.

	Start	**Finish**
1st Floor (east side) Rough-in	3/2/78	3/4/78
1st Floor (west side) Rough-in	3/5/78	3/8/78
2nd Floor (east side) Rough-in	3/22/78	3/24/78
3rd Floor (west side) Rough-in	3/25/78	3/29/78

The electrical inspector will begin scheduling his time by obtaining the estimated start and completion dates from either the general or the electrical contractor; it would be a good idea to obtain these schedules from both. These dates are then entered on a job progress schedule as shown in Fig. 13–2. Such a chart gives a graphic picture of the overall job from a calendar performance standpoint. The inspector should further enter the start and completion dates of all phases of the electrical work. If the electrical contractor cannot furnish this information, the inspector will have to estimate the conditions himself.

CALENDAR BAR CHART FOR ELECTRICAL PROJECTS

DATE PROJECT BEGAN 6 Feb 78 ; DATE COMPLETED 31 Aug 78 ; PREPARED BY J.E.T. DATE 9 Jan 78 PROJECT NO. 1385 PROJECT NAME P/F Bldg.

PERIOD	1	2	3	4	5	6	7	8	9	10	11	12	13	14	15	16	17	18	19	20	21	22	23	24	25	26	27	28	29
PERIOD END.	Feb 6	13	20	27	Mar 6	13	20	27	Apr 3	10	17	24	May 1	8	15	22	29	Jun 5	12	19	26	Jul 10	17	24	31	Aug 7	14	21	28

Periods 30 through 50 have no period-end dates.

	DESCRIPTION	code	ESTIMATE OF LABOR MAN-HOURS
1	ELECTRICAL INSTALLATION PLANNING + SCHEDULING		
2	On site - plan study Temporary Service	OPS TS	80 24
3	Branch rough-in	BR	1000
4	Feeder rough-in	FR	200
5	Panels and SW. Bd. Equip.	P	620
6	Branch wire pull	WP	280
7	Feeder wire pull	FP	100
8	Wiring devices	D	70
9	Lighting fixtures	F	1050
10	Special suspended fixture display bd.	FS	300
11	Equipment connect	EQ	40
12	Final cleanup + insp.	LC	40
13			
14			
15			
16			
17			
18			
19			
20			
21			
22			
23			
24			
	TOTALS		3804

FIG. 13-1. A typical bar graph.

JOB PROGRESS REPORT

JOB ________ Contract Amount ________ Sheet No. ____ of ____ Sheets

JOB NO. ________ Change Orders ________ Date ________

Dating	MATERIAL		LABOR		DIRECT JOB COST				MISCELLANEOUS		TOTAL COSTS		MAN HOURS	
	Monthly	Total	Monthly	Total	Monthly	Total	Monthly	Total	Monthly	Total	Monthly	Total	Monthly	Total

Form E-17 MINNESOTA ELECTRICAL ASS'N. 3100 Humboldt Ave. S. Minneapolis, Minn. 55408 CODE 1. INSURANCE 2. BONDS 3. PERMITS 4. TOOL COST 5. TRANSPORTATION 6. ________

FIG. 13-2. Job progress schedule.

On the chart, tentatively lay out the calendar durations for each section of the electrical work as related to required inspections. In doing so, it will be noted that some sections of the electrical work may not occur for the full duration of a given section of the building construction. For example, during the site layout and excavation, the electrical contractor will have little, if any, work on the project. However, the electrical inspector should check the drawings and specifications to determine the extent of any underground wiring. During the concrete construction of the footings, basement walls, etc., there will still be little electrical work; perhaps the insertion of a few conduit sleeves, for example—nothing the electrical inspector would be pushed on. Then at a certain point, the electrical work will pick up momentum and often extends over several sections of the general building construction. It is at this time that the inspector must plan his time carefully to be sure that work will not be held up due to his tardiness in getting to the job site for the inspections.

This initial planning and scheduling will probably have to be adjusted somewhat during the actual construction, but when working under such a plan, the inspector's time can be budgeted more efficiently, keeping work progress delays to a minimum.

The calendar bar chart just described provides the electrical inspector with a visual picture of the proposed job progress; it does not provide for the sections of the electrical work that have been inspected and approved. Therefore, a comparative bar chart must be provided to show the actual work inspected and approved on each phase of the project. Colored marks may be used more effectively than the solid bar of one color. One color, for example, can be used to denote the estimated job progress. Another color can be used to denote the work inspected. A third color may be used along side this last color to denote the findings of the inspection, that is, green for approved or red for turned down and requiring re-inspection.

The main point to remember when using a bar chart is to make certain that it is kept up-to-date and frequently referred to in order that the electrical inspectors have a continual picture of the progress of the job.

Once the electrical inspector gets into the habit of regularly checking the progress charts, any problems that occur will usually come to light immediately.

13.2 NETWORK DIAGRAMS

In the 1950s programs were undertaken to apply computers to construction scheduling. Initially the bar graph was used, but its inability to carry forward the intent of the original planner was soon recognized. However, by connecting the related activities with vertical lines, planners were able to show the interdependencies. Eventually, the time scale was dropped and the logic interconnected diagram or network was drawn in a time flow, but not to an exact scale.

This network alone is not sufficient to provide the time dimensions; estimates must be added for working time to each activity. The planning is however initially independent of the dates, and aims at determining feasibility. A typical example is shown in Fig. 13–3.

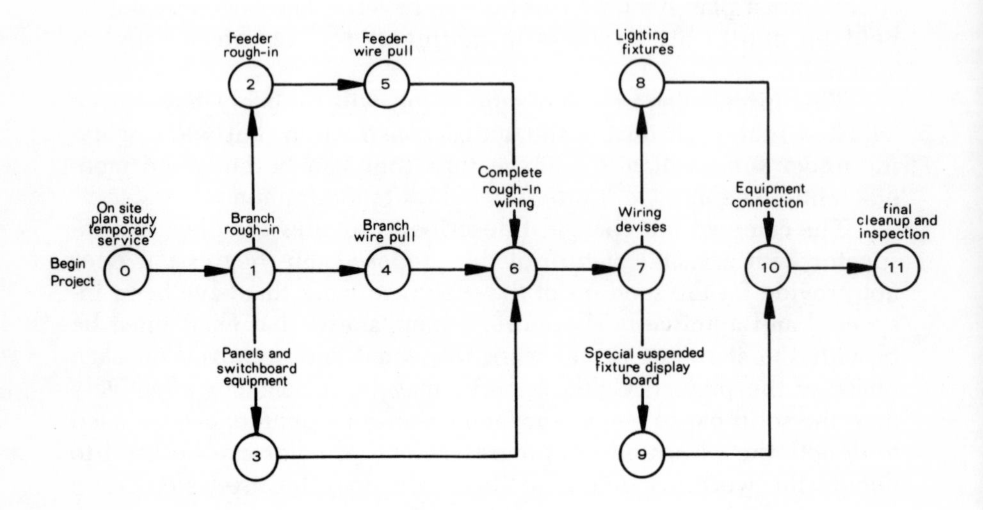

FIG. 13-3. A typical network diagram.

13.3 SPECIFICATIONS

The requirements for construction scheduling are usually included in the General Conditions of the written specifications. The requirements from the specifications for one project read as follows:

> *The contractor, immediately after being awarded the contract, shall prepare and submit for the Architect's approval an unlimited progress schedule for the work. The progress schedule shall be related to the entire project to the extent required by the contract documents. This sec-*

tion shall indicate the dates for the starting and completion of the various stages of construction and shall be revised as required by the conditions of the work, subject to the architect's approval.

Once the contractor has completed the progress schedule as outlined in the paragraph above, the electrical inspector can then take this schedule and tentatively prepare an inspection schedule of his own, so as not to hold up the progress of the job in any way.

Article 8—TIME—from the American Institute of Architects *General Conditions of the Contract for Construction* reads as follows:

ARTICLE 8
TIME

8.1 DEFINITIONS

8.1.1 The Contract Time is the period of time allotted in the Contract Documents for completion of the Work.

8.1.2 The date of commencement of the Work is the date established in a notice to proceed. If there is no notice to proceed, it shall be the date of the Agreement or such other date as may be established therein.

8.1.3 The Date of Substantial Completion of the Work or designated portion thereof is the Date certified by the Architect when construction is sufficiently complete, in accordance with the Contract Documents, so the Owner may occupy the Work or designated portion thereof for the use for which it is intended.

8.2 PROGRESS AND COMPLETION

8.2.1 All time limits stated in the Contract Documents are of the essence of the Contract.

8.2.2 The Contractor shall begin the Work on the date of commencement as defined in Subparagraph 8.1.2. He shall carry the Work forward expeditiously with adequate forces and shall complete it within the Contract Time.

8.3 DELAYS AND EXTENSIONS OF TIME

8.3.1 If the Contractor is delayed at any time in the progress of the Work by any act or neglect of the Owner or the Architect, or by any employee of either, or by any separate contractor employed by the

Owner, or by changes ordered in the Work, or by labor disputes, fire, unusual delay in transportation, unavoidable casualties or any causes beyond the Contractor's control, or by delay authorized by the Owner pending arbitration, or by any cause which the Architect determines may justify the delay, then the Contract Time shall be extended by Change Order for such reasonable time as the Architect may determine.

8.3.2 All claims for extension of time shall be made in writing to the Architect no more than fifteen days after the occurrence of the delay; otherwise they shall be waived. In the case of a continuing cause of delay only one claim is necessary.

8.3.3 If no schedule or agreement is made stating the dates upon which written interpretations as set forth in Subparagraph 1.2.5 shall be furnished, then no claim for delay shall be allowed on account of failure to furnish such interpretations until fifteen days after demand is made for the delay and not then unless such claim is reasonable.

8.3.4 This Paragraph 8.3 does not exclude the recovery of damages for delay by either party under other provisions of the Contract Documents.

Chapter 14
Change Orders

Change orders are legal documents describing and approving changes in conditions or scope to construction documents, including the contract signed by the owners and contractor. There are four basic types of change orders:

1. Additional or deleted work

2. Change in location without materially changing the extent of work

3. Change in the basis of performing the work

4. Changes in scope of work responsibility

In no case should the electrical inspector allow a change to be undertaken (if within the inspector's realm of responsibility) without written authorization from a responsible authority representing the owner or awarding authority.

14.1 PROCEDURE FOR CHANGE ORDERS

It is good management to use a specific procedure and related forms for authorizing such changes. For example, when a major change is made involving an additional contract between the electrical contractor and general contractor or awarding authority, a new work order or change order must be initiated. When any other than very minor changes are made involving a price adjustment to the existing contract, some form of change order is initiated stating that both parties agree to the change. Even when minor changes are made, which might not increase the price of the original contract, some form of written agreement should be issued. Then if a problem arises from the change, the blame can be put on the responsible party.

The following forms are examples of change orders that may be purchased for electrical construction work. A brief description of each is given.

The form in Fig. 14-1 is available from the Minnesota Electrical Association, Inc., 3100 Humboldt Avenue South, Minneapolis, Minn. 55408. It is $5\frac{1}{2}$ in. $\times$ $8\frac{1}{2}$ in. and is quite sufficient for most small to medium sized contracts. The form should be written in triplicate—providing one copy for the inspector's files, one for the contractor, and one for the owner/engineer/architect.

CHANGE ORDER

TO: .... Number ....
.... Date ....
.... Phone ....
Job Name/Location: Job No. ....
.... Date of Existing Contract:
....

****

....
....
....
....
....
....
....
....
....
....

NOTE: This revision becomes part of and in conformance with the existing contract.

****

WE AGREE hereby to make the change(s) specified above
at this price $ ....
Date .... Previous Contract Amount $ ....
Authorized Signature: Revised Contract Total $ ....
.... Date of Acceptance ....
(contractor) Signature ....
(owner)

FIG. 14-1. This small change order form is quite sufficient for most small to medium size contracts.

Another similar form for extra work is shown in Fig. 14-2. This form is primarily designed for on-the-job use by the contractor or superintendent to obtain a written and signed authorization for extra work or changes ordered on the job by the engineer, general con-

CHANGE ORDER

To.................................... Change Order No....................

Address.................................Date ....

.... Phone ....

Job Name/Location.......................Job No............................

.... Date of Existing Contract ....

****

....
....
....
....
....
....
....
....
....
....
....
....
....
....

NOTE: This revision becomes part of and in conformance with the existing contract

****

WE AGREE hereby to make the change(s) specified above at this price: $........

Date .... **Previous Contract Amount:** $........

Authorized Signature (Contractor) ....

Revised Contract Total: $........

Date of Acceptance............................

Signature....................................
(owner)

FIG. 14-2. A change order form primarily designed for on-the-job use.

tractor, or other contractors. Its use avoids future disclaiming of responsibility for having ordered the work or change. Space is provided for a brief description of the extra work and authorization and billing information. It is printed in triplicate providing copies for the inspector, the contractor, and the owner or his representative.

The form in Fig. 14-3 is designed to meet the requirements found on most contract forms that extra work or changes be covered by written orders. Space is provided for a written description of the extra work and authorization and billing information as well as general information pertaining to the extra work.

ORDER TO DO EXTRA WORK

DATE ____________

TO ____________
ELECTRICAL CONTRACTOR

YOU ARE HEREBY AUTHORIZED TO INSTALL THE FOLLOWING: ____________

WE AGREE TO PAY FOR THIS WORK AS FOLLOWS:

BASED ON ☐ ESTIMATED COST WHICH IS $______
☐ REGULAR TIME AND MATERIAL BASIS

IT IS AGREED THAT ANY COST ESTIMATE IS SUBJECT TO APPROVAL OF YOUR OFFICE.

ORDER TAKEN BY:

SIGNED ____________

BY ____________

DATE			LABOR-NAME OF WORKMAN	JOB NO.	HOURS

REMARKS

Form S-4

LIST MATERIAL USED FOR THIS JOB ON REVERSE SIDE

FIG. 14-3. This form is designed to meet the requirements found in construction documents stating that extra work or changes must be covered by written orders.

The Additional Work Authorization form in Fig. 14-4 provides a detailed record of all changes agreed upon, establishes a firm price and payment date for each change, and eliminates hasty and improper verbal agreements. As the name implies, this form should be used only for additional work, that is, work that requires additional materials and labor.

A Change Order form like the one in Fig. 14-5 may be used for any change in the original contract specifications. Such changes may include any of the following:

1. Additions or deletions of work from the original contract.

ADDITIONAL WORK AUTHORIZATION

CHANGE ORDER NO. . . . .

Owner's Name . . . . Phone . . . . Date . . . .

Street . . . . Job Number . . . .

City . . . . State . . . . Job Name . . . .

Existing Contract No. . . . . Street . . . .

Date of Existing Contract . . . . City . . . . State . . . .

* *

. . . .

. . . .

. . . .

. . . .

. . . .

. . . .

. . . .

. . . .

. . . .

. . . .

. . . .

. . . .

. . . .

ADDITIONAL CHARGE FOR ABOVE WORK IS: $ . . . .

Payment will be made as follows: . . . .

. . . .

Above additional work to be performed under same conditions as specified in original contract unless otherwise stipulated.

Authorized Signature . . . . Date . . . .
(owner)

* *

We hereby agree to furnish labor and materials – complete in accordance with the above specifications at above stated price.

Authorized Signature . . . . Date . . . .
(contractor)

FIG. 14-4. Additional work authorization form.

2. Changes in materials, method, or equipment.

3. Changing conditions due to strikes, weather, or other conditions not previously determined.

4. Changes due to omissions in the contract—omitted by the engineer/designer or the contractor.

5. Time extensions. Changes do cause delays and the time extension should be entered on a form such as the one in Fig. 14-5.

When several small changes are anticipated on a project, handy Change Order forms like the one shown in Fig. 14-6 should be

EXTRA WORK

DESCRIPTION ____________________

OWNER'S SIGNATURE ____________________

(CUSTOMER)

WORK LEFT TO BE DONE ____________________

FIG. 14-5. A change order form suitable for almost any change in the original contract specifications.

available on the job site to obtain authorized signatures for every job change. They are small and easy to carry. Such a form can prevent misunderstandings by eliminating hasty, improper verbal agreements.

14.2 USING THE CHANGE ORDER FORM

The following paragraphs describe the recommendations for carrying out changes in the work from the original construction documents.

On most projects, the owner or his representatives may order changes in the work within the general scope of the contract consisting of additions, deletions, or other revisions. In doing so, however, the contract sum and the contract time must be adjusted accordingly. All such changes in the work must be authorized by a Change Order, and must be executed under the applicable conditions of the contract documents.

CUSTOMER'S ORDER NO.

WORK ORDER

OUR JOB NUMBER

DATE ______ 19__

ORDERED BY ______ POSITION ______

COMPANY ______

ADDRESS ______

CHARGE TO ______

ORDER. REC'D BY ______

PHONE ___ MAIL ___ PERSONAL ___

CONFIRMED ______

LOCATION OF JOB

INSTRUCTIONS: WORK TO BE DONE:

JOB ASSIGNED TO ______

CUSTOMER'S ACKNOWLEDGEMENT:

THIS WORK HAS BEEN COMPLETED SATISFACTORILY.

SIGNED ______ CUSTOMER

DATE ______ BY ______

Form S-5

FIG. 14-6. When several small changes are anticipated on a project, handy change order forms like the one shown here, should be available on the job site.

A Change Order, as described previously, is a written order to the contractor signed by the owner, architect, or other authorized person, issued after the execution of the original contract, authorizing a change in the work or an adjustment in the contract sum or the contract time. Alternatively, the change order may be signed by the head of the inspection team alone, provided he has written authority from the owner for such procedure and that a copy of such written authority is furnished to the contractor upon request. The contract sum and the contract time may be changed only by a properly prepared Change Order.

The cost or credit to the owner resulting from a change in the work must be determined in one or more of the following ways.

1. By mutual acceptance of a lump sum properly itemized.

2. By unit prices stated in the contract documents or subsequently agreed upon.

3. By cost and a mutually acceptable fixed or percentage fee.

If none of the above methods can be agreed upon, the contractor, provided he received a written change order, should promptly proceed with the work involved so as not to delay the project any more than necessary. The cost of such work will then be determined by the architect or some other authorized person on the basis of the contractor's reasonable expenditures and savings, including, in the case of an increase in the contract sum, a reasonable allowance for overhead and profit. In such case, the contractor must keep and present, in such form as the inspector or architect may prescribe, an itemized accounting together with appropriate supporting data. Pending final determination of cost to the owner, payments on account shall be made on the Architect's Certificate for Payment. The amount of credit to be allowed by the contractor to the owner for any deletion or change resulting in a net decrease in cost will be the amount of the actual net decrease as confirmed by the architect. When both additions and credits are involved in any one change, the allowance for overhead and profit shall be figured on the basis of net increase, if any.

If unit prices are stated in the contract documents or subsequently agreed upon, and if the quantities originally contemplated are so changed in a proposed Change Order that application of the agreed unit prices to the quantities of work proposed will create a hardship on the owner or the contractor, the applicable unit prices must be equitably adjusted to prevent such hardships.

Should concealed conditions encountered in the performance of the work below the surface of the ground be at variance with the conditions indicated by the contract documents, or should unknown physical conditions below the surface of the ground of an unusual nature be encountered, differing materially from those ordinarily encountered and generally recognized as inherent in work of the character provided for in the contract, the contract sum should be equitably adjusted by Change Order upon claim by either party made within a reasonable time after the first observance of the condition.

If the contractor wishes to make a claim for an increase in the contract sum or an extension in the contract time, he shall give the architect written notice thereof within a reasonable time after the occurrence of the event giving rise to such claim. This notice must be given by the contractor before proceeding to execute the work, except in an emergency endangering life or property, in which case the contractor will proceed in accordance with the procedures set forth in the contract documents. No such claim should be valid unless done in accordance with procedures set forth in the contract documents. If the owner and the contractor cannot agree on the amount of the adjustment in the contract sum or the contract time, it will be determined by the architect. Any change in the contract sum or contract time resulting from such claim must be authorized by Change Order.

The architect or the architect's inspection team usually has the authority to order minor changes in the work not involving an adjustment in the contract sum or an extension of the contract time, and not inconsistent with the intent of the contract documents. Such changes may be effected by field orders or by other written notice. Such changes are binding on the owner and the contractor.

The architect may issue written field orders that interpret the contract documents in accordance with the written specifications or that order minor changes in the work in accordance with the above paragraph without change in the contract sum or contract time. The contractor should carry out such field orders promptly.

14.3 SUMMARY

Nowadays it is a rare electrical project of any size that does not undergo some change from the original contract prior to the completion of the project. Adding to the original contract is quite common; changing or deleting certain portions of the work that require additional charges to be made or credits given to the customer occurs frequently. When representing the owner, the electrical inspector must make certain that no changes in the work occur without an authorization signed by someone in authority. When a change in the contract price is involved, a later dispute can be avoided when a signed Change Order can be produced, stating exactly what changes were to have taken place. It is also a good policy to have a signed authorization for other than very minor location changes *even though* no change in price is involved. This can help settle any later disputes concerning appropriateness of making the change.

Electrical inspectors working for architectural or engineering firms are usually directly responsible for handling such changes in the proper manner when they are of an on-the-job instruction basis, and not formally handled by the contractor and the architect directly—with formal change order instructions initiated in the architect's office.

To facilitate on-the-job handling of such change authorization, the electrical inspector should be furnished with appropriate change order forms he can fill out on the job and have signed by himself (if authorized) and the contractor performing the changes. Since additions to the job are involved in a majority of instances, these forms are usually referred to as Extra Work Orders, Change Orders, etc., and are shown in Figs. 14-1 through 14-6.

Some forms are used mainly to show authorization of the extra work or change; and sometimes the price to be charged or credit allowed is usually negotiated by the architect/engineer/owner and the contractor. It is important, however, for the inspector to ascertain who is to be billed for the work and the basis of billing—that is, cost-plus, time, and material; or if a lump-sum price is to be established. Such information should be entered on the forms at the time it is signed by the contractor and the person authorizing the work or change.

Chapter 15
Progress Payments

One of the most important functions of the electrical inspector employed by architectural/engineering firms is the approval of progress payments. This is a very important responsibility for the inspector; the contractor needs to recover his invested money as quickly as possible, and the inspector should not delay in handling this part of his job. If problems exist, call them to the contractor's attention immediately and make adjustments so that the contractor can obtain a certain amount of his request—even though it is not the full amount. Many contractors have been crippled tremendously because of slow progress payments.

The basis of progress payments is an estimate of the work actually accomplished. The progress is generally measured by a breakdown of the various portions of the electrical system presented by the contractor's estimator before construction begins. A certain portion of the contract price is allotted for rough-in wiring on each floor, a certain amount for feeders, panelboards, fixtures, and so on. With this schedule, the inspector can determine the dollar value of work performed by comparing the schedule to actual field observations of the work, by acceptance of contractor statements, or a combination of the two. The exact method of verifying progress payment requests varies according to the type of contract. There are many variations of construction contracts, but most break down into the following:

In the majority of electrical construction work, the general contractor will request quotations on a given project from all subcontractors and award the job to those submitting the lowest quotation or bid. In some instances, the general contractor (or the owner if the electrical contractor is hired directly) may not solicit competitive quotations from the electrical contractor, but award the work to a competent electrical contractor on a firm-price basis, a total-cost-plus-a-percentage basis (cost-plus), or a total-cost-plus-a-predetermined-fixed-fee basis.

Regardless of the method of awarding the contract, after it is awarded, the electrical contractor is required to furnish a breakdown of his total contract price into several segregated items such as large and small conduit rough-in, wire and cable, panelboards and pull boxes, wiring devices, lighting fixtures, lamps, and special equipment. As mentioned previously, such price breakdowns are usually required to assist in the preparation of monthly job progress reports and payments.

15.1 MONTHLY PROGRESS REPORTS AND PAYMENTS

Each month the electrical inspector will make an estimate of the percentage of the work completed at that time for each of the segregated groups on the job price breakdown. The percentage of completion shown on the progress report for the previous month is deducted from the current monthly progress report, and the difference indicates the percentage of work completed for each group of work during the last month.

In calculating the value of the progress payment to be made for a given month's work, the total value of all the segregated groups of work performed up to the end of that month is determined, from which amount the total for the previous month is deducted. The difference is the value of the work done during the month. In most cases, a retainage of ten to twenty percent is held out of the payment; this is ordinarily paid thirty days after the completion of the project.

In many cases, the contract documents allow the electrical contractor to be paid for the value of material delivered to the job but not yet installed, provided the material is stored properly and secured under lock and key.

15.2 TYPICAL PROGRESS PAYMENT PROCEDURES

Before the first application for payment, the contractor must submit to the architect's representative a schedule of values of the various portions of the work, including quantities if required by the architect, aggregating the total contract sum, divided to facilitate payments. This report should be submitted on a form as specified or as the architect and the contractor may agree upon, and supported by such data to substantiate its correctness. Each item in the schedule of values must include its proper share of overhead and profit. This schedule, when approved by the architect, is used only as a basis for the contractor's application for payment.

At least ten days before each progress payment is due, the contractor must submit to the architect an itemized application for payment, supported by such data substantiating the contractor's right to payment as the owner or architect may require.

If payments are to be made on account of materials or equipment not incorporated in the work but delivered and suitably stored at the site, or at some other location agreed upon in writing, such payments will be conditioned upon submission by the contractor of bills of sale or such other procedures satisfactory to the owner to establish the owner's title to such materials or equipment or otherwise protect the owner's interest including applicable insurance and transportation to the site.

The contractor warrants and guarantees that title to all work, materials, and equipment covered by an application for payment, whether incorporated in the project or not, will pass to the owner upon the receipt of such payment by the contractor, free and clear of all liens, claims, security interests, or encumbrances, referred to "liens"; and that no work, materials, or equipment covered by an application for payment will have been acquired by the contractor; or by any other person performing the work at the site or furnishing materials and equipment for the project, subject to an agreement under which an interest therein or an encumbrance thereon is retained by the seller or otherwise imposed by the contractor or such other person.

If the contractor has made application for payment as above, the architect will, with reasonable promptness but not more than seven days after the receipt of the application, issue a certificate for payment to the owner, with a copy to the contractor, for such amount as he determines to be properly due, or state in writing his reasons for withholding a certificate.

The issuance of a certificate for payment will constitute a representation by the architect to the owner, based on his observations at the site and the data comprising the application for payment, that the work has progressed to the point indicated; that, to the best of his knowledge, information, and belief, the quality of the work is in accordance with the contract documents (subject to an evaluation of the work as a functioning whole upon substantial completion, to the results of any subsequent tests required by the contract documents correctable prior to completion, and to any specific qualifications stated in his certificate); and that the contractor is entitled to payment in the amount certified. In addition, the architect's final certificate for payment will constitute a further representation that the conditions precedent to the contractor's being entitled to final payment have been fulfilled. However, by issuing a certificate for payment, the architect shall not thereby be deemed to represent that he has made exhaustive or continuous on-site inspections to check the quality or quantity of the work or that he has reviewed the construction means, methods, techniques, sequences, or procedures, or that he has made any examination to ascertain how or for what purpose the contractor has used the moneys previously paid on account of the contract sum.

After the architect issues a certificate for payment, the owner shall make payment in the manner provided in the agreement.

No certificate for a progress payment, nor any progress payment, nor any partial or entire use or occupancy of the project by the owner shall constitute an acceptance of any work not in accordance with the contract documents.

The architect may decline to approve an application for payment and may withhold his certificate in whole or in part if in his opinion he is unable to make representations to the owner. The architect may also decline to approve any applications for payment or, because of subsequently discovered evidence or subsequent inspections, he may nullify the whole or any part of any certificate for payment previously issued to such extent as may be necessary in his opinion to protect the owner from loss because of:

1. defective work not remedied;
2. claims filed or reasonable evidence indicating probable filing of claims;
3. failure of the contractor to make payments properly to subcontractors or for labor, materials, or equipment;
4. reasonable doubt that the work can be completed for the unpaid balance of the contract sum;
5. damage to another contractor;
6. reasonable indication that the work will not be completed within the contract time; or
7. unsatisfactory prosecution of the work by the contractor.

When the above grounds are removed, payment shall be made for amounts withheld because of them.

If the architect should fail to issue any certificate for payment, through no fault of the contractor, within seven days after receipt of the contractor's application for payment, or if the owner should fail to pay the contractor within seven days after the date of payment established in the agreement any amount certified by the architect or awarded by arbitration, then the contractor may, upon seven additional days' written notice to the owner and the architect, stop the work until payment of the amount owing has been received.

When the contractor determines that the work or a designated portion thereof acceptable to the owner is substantially complete, the contractor shall prepare for submission to the architect a list of items to be completed or corrected. The failure to include any items on such list does not alter the responsibility of the contractor to complete all work in accordance with the contract documents. When the architect on the basis of an inspection determines that the work is substantailly complete, he will then prepare a certificate of substantial completion, shall state the responsibilities of the owner and the contractor for maintenance, heat, utilities, and insurance, and shall fix the time within which the contractor shall complete the items listed therein, said time to be within the contract time. The certificate of substantial completion shall be submitted to the owner and the contractor for their written acceptance of the responsibilities assigned to them in such certificate.

Upon receipt of written notice that the work is ready for final inspection and acceptance and upon receipt of a final application for payment, the architect will promptly make such inspection and, when he finds the work acceptable under the contract documents and the contract fully performed, he will promptly issue a final certificate for payment stating that to the best of his knowledge, information, and belief, and on the basis of his observations and inspections, the work has been completed in accordance with the terms and conditions of the contract documents and that the entire balance found to be due the contractor, and noted in said final certificate, is due and payable.

Neither the final payment nor the remaining retained percentage shall become due until the contractor submits to the architect (1) an affidavit that all payrolls, bills for materials and equipment, and other indebtedness connected with the work for which the owner or his property might in any way be responsible, have been paid or otherwise satisfied; (2) consent of surety, if any, to final payment; and (3) if required by the owner, other data establishing payment or satisfaction of all such obligations, such as receipts, releases, and waivers of liens arising out of the contract, to the extent and in such form as may be designated by the owner. If any subcontractor refuses to furnish a release or waiver required by the owner, the contractor may furnish a bond satisfactory to the owner to indemnify him against any such lien. If any such lien remains unsatisfied after all payments are made, the contractor shall refund to the owner all moneys that the latter may be compelled to pay in discharging such lien, including all costs and reasonable attorney's fees.

If after substantial completion of the work, final completion thereof is materially delayed through no fault of the contractor, and the architect so confirms, the owner shall, upon certification by the architect, and without terminating the contract, make payment of the balance due for that portion of the work fully completed and accepted. If the remaining balance for work not fully completed or corrected is less than the retainage stipulated in the agreement, and if bonds have been furnished, the written consent of the surety to the payment of the balance due for that portion of the work fully completed and accepted shall be made under the terms and conditions governing final payment, except that it shall not constitute a waiver of claims.

The making of final payment shall constitute a waiver of all claims by the owner except those arising from:

1. unsettled liens,

2. faulty or defective work appearing after substantial completion,

3. failure of the work to comply with the requirements of the contract documents, or

4. terms of any special guarantees required by the contract documents.

The acceptance of final payment shall constitute a waiver of all claims by the contractor except those previously made in writing and still unsettled.

Chapter 16
Documentation

The electrical inspector should keep and have job records of every significant activity occurring throughout the project. Records should include the dates and condition of the rough wiring for example. Records of any faults and when faults are corrected should be made. All change orders, especially, should be kept in order and available at all times. Some inspectors are required to keep work logs, listing every significant step of the project.

The purpose of good job records covers a wide variety of instances where documented facts are essential. Job mishaps, accidents, preparation of claims, defense against claims, etc., are all quite common on every electrical project of any size. Therefore, a documentation plan insures that a comprehensive set of documentation is maintained, and that record keeping is cut to the bare minimum.

16.1 COMPLETING THE PROJECT

After an electrical installation has been substantially completed, there will usually be several areas of the project not quite ready to be turned over to the owners. There are many reasons for this situation:

1. The building site may have to be graded before the electrical contractor can install certain outside electrical work, such as lighting standards and fixtures for parking lots, decorative lighting, etc.

2. The electrical contractor may have difficulty getting delivery of a few specialized pieces of electrical equipment.

3. Another trade may have difficulty in getting delivery on certain pieces of equipment that require an electrical connection by the electrical contractor.

Obviously, at this point, the owners of the project are going to be anxious to move in and the electrical inspector should make every effort to push the electrical contractor in getting his portion of the project completed. In fact, near the completion of any project, the electrical inspector should arrange a meeting with the electrical contractor to discuss the project step-by-step to see if there are going to be any problems with material deliveries or other trades in finishing the project. If any problems are foreseen, immediate action should be taken. A record will be made of all unfinished areas and the completion of them should be followed through in a methodical manner until the problems are solved and the electrical work is completed.

When the electrical contractor feels that the project is finished, he will call for a final inspection so that he can be released from the project and obtain the balance of his money. When this request is made, the electrical inspector (whether working for an architectural/engineering firm or municipal inspection department) should inspect the project as soon as possible. Then if there are any discrepancies, notice may be given and the electrical contractor will have ample time to correct them without holding up the general contractor or other trades involved in the project.

16.1.1 Punch Lists

In some cases the owner will insist on moving in a building prior to its completion; he is willing to have small deficiencies corrected later. When such a condition exists, the electrical inspector and the electrical contractor should make a room-by-room and floor-by-floor list of all items still missing or wiring methods that must be corrected.

This list, often called a "punch list," should include all items, regardless of how minor, that the electrical inspector detects. Once completed, the project should now be ready for a preliminary inspection by the various inspection teams.

During the inspection, the electrical inspector should have the electrical contractor accompany him on his tour to be sure that all criticisms are understood. Occasionally the owner of a project will ask for work beyond the scope of the contract documents, or for items that the contractor may think are beyond the scope of the contract. This is where documentation of all items pays off.

Any items that present a problem should be listed on this inspection tour. Do not stop to check out each one as the tour progresses. Rather, make the list so that they may be checked against the contract documents once the tour is completed. Then the inspection team can go over the listed projects one by one until they are straightened out.

If all complaints by the inspection team are valid, a punch list should be forwarded to the electrical contractor within a reasonable time after the inspection. Copies should also be sent to the consulting engineer who designed the electrical system, the architect, and the owners.

If at all possible, the electrical inspector should see that the items on the punch list are taken care of before the owners begin moving in. There are several valid reasons for this suggestion:

1. The owner's workmen will be moving in materials, placing furniture (possibly over or near some punch-list items), and organizing the interior space in general.

2. If there is damage to any of the electrical equipment prior to the owner moving in, there is no question as to who pays for it. However, if one of the owner's crew breaks the electrical equipment, blame may be difficult to prove, and someone not responsible may end up paying for the damage.

After all the corrections requested in the preliminary punch list have been completed, the electrical contractor will request a final inspection by the local building authorities, who will send out one or more electrical inspectors to check the quality of workmanship and to insure that all work complies with the NEC and all local ordinances.

16.2 TESTING THE ELECTRICAL SYSTEM FOR FAULTS

With all the corrections accomplished, the electrical inspector will not have to make another detailed inspection. However, many contracts call for a final testing of the electrical system. These tests should be made in the presence of the electrical inspector. The extent of this test can vary from job to job, but the following is typical of most electrical specifications:

1. The electrical contractor shall take certain voltage and current readings, record all values, and submit the reading in triplicate

to the engineer or his electrical inspector. Two complete sets of readings are required, one under *no load* and one under *maximum available load.* The current and voltage shall be recorded on each phase (plus voltage between phases) at main panelboard and at each branch circuit panelboard. Additional spot readings shall be made if required by the electrical inspector in charge of the project. Resistance of grounding conductors shall be tested and recorded. Forms for submitting this report may be obtained from the inspector's office.

2. The electrical contractor shall also take voltage readings on each phase of each motor circuit and each resistance heater circuit installed under the contract; these readings must be recorded as described previously. The electrical contractor shall also report motor nameplate data, actual motor heater protective device, and all other data necessary for selection of heater device.

The requirements of the previous specifications may seem rigid to those electrical inspectors who inspect mostly residential and small commercial projects; for those involved in industrial installations perhaps it seems lenient. The requirements of an electrical specification for a typical motel reads as follows:

1. It shall be the responsibility of the electrical contractor to connect the electrical loads to provide minimum phase unbalance throughout the building. The electrical contractor shall operate the building under full heating and other load conditions, with full lighting and provide a record of the amperage per phase for each feeder installed to the main distribution panel.

2. As soon as electric power is available and connected to serve the equipment in the building, and everything is ready for final testing and placing in service, a complete operational test shall be made. The electrical contractor shall furnish all necessary instruments and testing equipment to make all tests, adjustments, and trial operations required to place the system in balanced and satisfactory operational condition. He shall further furnish all necessary assistance and instructions to properly instruct the owner's authorized personnel in the operation and care of the system.

3. Prior to testing the system, the feeders and branch circuits shall be continuous from main feeders to main panels, to subpanels, to outlets, with all circuit breakers and fuses in place. The system shall be tested free from ground faults. Such tests shall be made in the presence of the electrical inspector.

4. No circuits shall be energized without the owner's approval.

16.3 DEMONSTRATION OF COMPLETE ELECTRICAL SYSTEM

To summarize the demonstration of a complete electrical system, the following specifications appeared in recent construction documents:

1. The electrical contractor shall provide the owner with a certification of the inspection and approval of an active member of the International Association of Electrical Inspectors of all work completed and included in the section, if required. The electrical contractor shall be responsible for notifying the inspector when work reaches inspection stage.

2. The electrical contractor shall be responsible for notifying the local authority having jurisdiction in order that local inspection may be carried out at the proper stages of construction.

3. The electrical contractor is responsible for paying all costs including permits, inspection fees, and any installation fees as required to complete the work under the electrical contract.

4. The electrical contractor shall guarantee the materials and workmanship for a period of 12 months from the time the installation is accepted by the owners. If, during this time, any defects should occur due to any defective materials, workmanship, negligence, or want of proper care on the part of the electrical contractor, he shall furnish any new materials as necessary, repair said defects, and put the system in order at his own expense on receipt of notice of such defects from the architect or his electrical inspector. This specification is not intended to imply that the electrical contractor shall be responsible for negligence of the owner.

16.4 IDENTIFICATION

The electrical inspector should require the electrical contractor to maintain accurate records of all deviations in work as actually installed from work indicated on the working drawings and as specified in the written specifications. Two complete sets of marked-up drawings should be delivered to the inspector to keep in the architect's files.

As mentioned in an earlier chapter, all panelboard directories should be neatly typed listing all equipment, circuits, and so forth, that each circuit breaker controls (see Fig. 6–9). Furthermore, the electrical inspector should require that either laminated plastic or rigid phenolic plastic nameplates be provided and mounted on all motor starters, disconnect switches, pushbutton stations, and power and lighting panels, and at each ceiling or wall access panel to electrical work where practical. The lettering should be ⅛-in. engraved letters and the plates should be attached with self-tapping sheet metal screws.

16.5 REQUISITIONING FINAL PAYMENTS

After the electrical contractor delivers to the owner, architect, or electrical inspector a certification of the approved electrical inspection by municipal inspectors and after the final inspection by the owner/architect's inspection team, a certificate should be issued immediately by the electrical inspector or architect authorizing final payment subject to the guarantees specified in the construction documents. If some items are still missing, including forms which have not been completed by the electrical contractor, the architect or electrical inspector may recommend withholding a small amount (usually 15% of the total contract price) to cover these items. At such times as the items are corrected, the retainage will be released to the electrical contractor.

A typical "Schedule of Values and Certificate of Payment" form is shown in Fig. 16–1. All such forms may differ slightly, but all will be basically the same. The left-hand column (1,2, etc.) gives the item number; the second column from the left is for a description of the job breakdown (branch circuit wiring, panelboards, feeders, etc.); the column to follow is for the original value of the work as given in the contractor's estimate to the inspector at the beginning of the project. If this is the final payment, obviously all items used on the very right-hand column should contain the 100% figure.

G. O. Form E & B CO-10 (Rev. 5-71)

SCHEDULE OF VALUES AND CERTIFICATE OF PAYMENT No. ______

____
Institution or Agency

Building or Project ______________________ Project No. ______

Contractor ______________________

For Period ______________ 19____ To ______________ 19____

Item No.	DESCRIPTION	ORIGINAL SCHEDULE OF VALUES	VALUE OF WORK COMPLETED			
			Previous Value	Value This Report	Total To Date	Percent Complete
1						
2						
3						
4						
5						
6						
7						
8						
9						
10						
11						
12						
13						
14						
15						
16						
17						
18						
19						
20						
21						
22						
23						
24						
25						
26						
27						
28						
29						
30						
31						
32						
33						
34						
35	TOTAL					
36	CHANGE ORDER-PLUS-MINUS					
37	ADJUSTED TOTAL					
38	LESS 10% RETAINED	XXXXXXXXXXXXXX				
39	NET TOTAL	XXXXXXXXXXXXXX				
40	AMT. THIS CERTIFICATION	XXXXXXXXXXXXXX	XXXXXXXXXXXXXX		XXXXXXXXXXXXXX	

Date ______________ 19____ Contractor ______________________

By ______________________

FIG. 16-1. Typical Schedule of Values and Certificate of Payment form.

Glossary

Accessible *(as applied to equipment)* Admitting close approach because equipment is not guarded by locked doors, elevation, or other effective means.

Accessible *(as applied to wiring methods)* Capable of being removed or exposed without damaging the building structure or finish, or not permanently closed in by the structure or finish of the building.

Aggregate Inert material mixed with cement and water to produce concrete.

Alternating current Alternating current (ac) is current that reverses direction rapidly, flowing back and forth in the system with regularity. This reversal of current is due to reversal of voltage occurring at the same frequency. In alternating current, any one wire is first positive, then negative, then positive, and so on.

Alternator An electric generator designed to supply alternating current. Some types have a revolving armature and in other types a revolving field.

Ampacity Current-carrying capacity expressed in amperes.

Ampere The ampere is the unit of measurement for electric current. It represents the rate at which current flows through a resistance of one ohm by a pressure of one volt.

Amplitude The maximum instantaneous value of an alternating voltage or current. It is measured in either the positive or negative direction.

Appliance Utilization equipment, generally equipment other than industrial, normally built in standardized sizes or types and installed or connected as a unit to perform one or more functions, such as clothes washing, air conditioning, food mixing, deep frying, etc.

Appliance, fixed An appliance that is fastened or otherwise secured at a specific location.

Appliance, portable An appliance that is actually moved or can easily be moved from one place to another in normal use.

Appliance, stationary An appliance that is not easily moved from one place to another in normal use.

Approved Acceptable to the authority enforcing the Code.

Attachment plug *(plug cap) (cap)* A device that, upon insertion in a receptable, establishes a connection between the conductors of the attached flexible cord and the conductors connected permanently to the receptacle.

Automatic Self-acting, operating by its own mechanism when actuated by some impersonal influence, such as a change in current strength, pressure, temperature, or mechanical configuration.

Backfill Loose earth placed outside foundation walls for filling and grading.

Bearing plate Steel plate placed under one end of a beam or truss for load distribution.

Bearing wall Wall supporting a load other than its own weight.

Bench mark Point of reference from which measurements are made.

Bonding jumper A reliable conductor used to ensure the required electrical conductivity between metal parts required to be electrically connected.

Branch circuit That portion of a wiring system extending beyond the final overcurrent device protecting the circuit.

Branch circuit, appliance A circuit supplying energy to one or more outlets to which appliances are connected; such circuits have no permanently connected lighting fixtures that are not a part of an appliance.

Branch circuit, general purpose A branch circuit that supplies a number of outlets for lighting and appliances.

Branch circuit, individual A branch circuit that supplies only one piece of utilization equipment.

Bridging System of bracing between floor beams to distribute floor load.

Building A structure that stands alone or that is cut off from adjoining structures by fire walls with all openings therein protected by approved fire doors.

Bus bar The heavy copper or aluminum bar used on switchboards to carry current.

Cabinet An enclosure designed for either surface or flush mounting and provided with a frame, mat, or trim in which swinging doors are hung.

Capacitor or condenser An electrical device that causes the current to lead the voltage, opposite in effect to inductive reactance. It is used to neutralize the objectional effect of lagging (inductive reactance), which overloads the power source.

Cavity wall Wall built of solid masonry units arranged to provide air space within the wall.

Chase Recess in inner face of masonry wall providing space for pipes and/or ducts.

Circuit breaker A device designed to open and close a circuit by nonautomatic means and to open the circuit automatically on a predetermined overload of current, without injury to itself when properly applied within its rating.

Coaxial cable A cable consisting of two conductors concentric with and insulated from each other. The braid is the outer conductor (often at ground) and shields the inner conductor.

Column Vertical load-carrying member of a structural frame.

Commutator Device used on electric motors or generators to maintain a uni-directional current.

Concealed Rendered inaccessible by the structure or finish of the building. Wires in concealed raceways are considered concealed, even though they may become accessible by withdrawing them.

Conductance The ability of material to carry an electric current.

Conductor Substances offering little resistance to the flow of electric current. Silver, copper, and aluminum are good conductors although no material is a perfect conductor.

Conductor, bare A conductor having no covering or insulation whatsoever.

Conductor, covered A conductor having one or more layers of non-conducting materials not recognized as insulation under the Code.

Conductor, insulated A conductor covered with material recognized as insulation.

Connector, pressure *(solderless)* A connector that establishes the connection between two or more conductors or between one or more conductors and a terminal by means of mechanical pressure and without the use of solder.

Continuous load A load in which the maximum current is expected to continue for three hours or more.

Contour line A line that is found on a land map denoting elevations and which connects points with the same elevation.

Controller A device, or group of devices, that serves to govern in some predetermined manner the electric power delivered to the apparatus to which it is connected.

Cooking unit, counter-mounted An assembly of one or more domestic surface heating elements for cooking purposes designed to be flush mounted in, or supported by, a counter and complete with internal wiring and inherent or separately mounted controls.

Crawl space Shallow space between the first tier of beams and the ground (no basement).

Current The flow of electricity in a circuit. It is expressed in amperes and represents an amount of electricity.

Curtain wall Nonbearing wall between piers or columns for the enclosure of the structure; not supported at each story.

Cycle One complete period of flow of alternating current in both directions. One cycle represents 360°.

Demand factor In any system or part of a system, the ratio of the maximum demand of the system, or part of the system, to the total connected load of the system, or part of the system, under consideration.

Direct current Current (dc) which flows in one direction only. One wire is always positive, the other negative.

Disconnecting means A device, a group of devices, or other means whereby the conductors of a circuit can be disconnected from their source of supply.

Double-strength glass Sheet glass that is one-eighth in. thick (single-strength glass is one-tenth in. thick).

Dry wall Interior wall construction consisting of plaster boards, wood paneling, or plywood nailed directly to the studs without application of plaster.

Duty, continuous A requirement of service that demands operation at a substantially constant load for an indefinitely long time.

Duty intermittent A requirement of service that demands operation for alternate intervals of (1) load and no load; (2) load and rest; or (3) load, no load, and rest.

Duty, periodic A type of intermittent duty in which the load conditions regularly recur.

Duty, short-time A requirement of service that demands operations at loads and for intervals of time, both of which may be subject to wide variation.

Electrical generator A machine so constructed that when its rotor is driven by an engine or other prime mover, a voltage is generated.

Electrode A conducting element used to emit, collect, or control electrons and ions.

Electron A particle of matter negatively charged.

Electron emission The release of electrons from the surface of a material into surrounding space due to heat, light, high voltage, or other causes.

Elevation Drawing showing the projection of a building on a vertical plane.

Enclosed Surrounded by a case that will prevent anyone from accidentally contacting live parts.

Equipment A general term including material, fittings, devices, appliances, fixtures, apparatus, and the like used as a part of, or in connection with, an electrical installation.

Expansion bolt Bolt with a casing arranged to wedge the bolt into a masonry wall to provide an anchorage.

Expansion joint Joint between two adjoining concrete members arranged to permit expansion and contraction with changes in temperature.

Exposed *(as applied to live parts)* Live parts that a person could inadvertently touch or approach nearer than a safe distance. This term is applied to parts not suitably guarded, isolated, or insulated.

Exposed *(as applied to wiring method)* Not concealed.

Externally operable Capable of being operated without exposing the operator to contact with live parts.

Facade Main front of a building.

Farad A unit of the charge capability of a capacitor.

Feedback The process of transferring energy from the output circuit of a device back to its input.

Feeder The conductors between the service equipment or the generator switchboard of an isolated plant, and the branch circuit overcurrent device.

Frequency Frequency of alternating current is the number of cycles per second. A 60-hertz alternating current makes 60 complete cycles of flow back and forth (120 alternations) per second.

Filter A combination of circuit elements designed to pass a definite range of frequencies, reducing all others.

Fire stop Incombustible filler material used to block interior draft spaces.

Fitting An accessory such as a locknut, bushing, or other part of wiring system that is intended primarily to perform a mechanical rather than an electrical function.

Flashing Strips of sheet metal bent into an angle between the roof and wall to make a watertight joint.

Footing Structural unit used to distribute loads to the bearing materials.

Frostline Deepest level below grade to which frost penetrates in a geographic area.

Fuse A protective device inserted in series with a circuit.

Gain The ratio of output to input power, voltage, or current, respectively.

Garage A building or portion of a building in which one or more self-propelled vehicles carrying volatile, flammable liquid for fuel or power are kept for use, sale, storage, rental, repair, etc.

Ground A conducting connection, whether intentional or accidental, between an electrical circuit or piece of equipment and earth or some other conducting body serving in place of the earth.

Grounded Connected to earth or to some conducting body that serves in place of the earth.

Grounded conductor A system or circuit conductor that is intentionally grounded.

Grounding conductor A conductor used to connect equipment or the grounded circuit of a wiring system to a grounding electrode.

Henry The basic unit of inductance.

Hertz A unit of frequency, one cycle per second. Written as 50-hertz or 60-hertz current, etc.

I-beam Rolled steel beam or built-up beam of I section.

Impedance Effects placed on alternating current by inductive capacitance (current lags voltage), capacitive reactance (current leads voltage), and resistance (opposes current but doesn't lag or lead voltage), or any combination of two. It is measured in ohms like resistance.

Incombustible material Material that will not ignite or actively support combustion in a surrounding temperature of 1200 °F during an exposure of 5 minutes; also material that will not melt when the temperature of the material is maintained at 900 °F for a period of at least 5 minutes.

Inductance The property of a circuit or two neighboring circuits that determines how much voltage will be induced in one circuit by a change of current in either circuit.

Inductor A coil.

Integrated circuit A circuit in which different types of devices such as resistors, capacitors, and transistors are made from solid-state components integrated into a tiny multi-circuit chip and connected to form a circuit.

Isolated Not readily accessible to persons unless special means for access are used.

Insulator Substances offering great resistance to the flow of electric current such as glass, porcelain, paper, cotton, enamel, and paraffin are called insulators because they are practically nonconducting. However, no material is a perfect insulator.

Jamb Upright member forming the side of a door or window opening.

KVA The abbreviation of kilovolt-amperes,which is the product of the volts times the amperes divided by 1000. This term is used in rating alternating current machinery because with alternating currents, the product of the volts times the amperes usually does not give the true average power.

KVAR The abbreviation of kilovolt-ampere reactance, which is a measurement of reactive power that generates power within induction equipment (motors, transformers, holding coils, lighting ballasts, etc.)

KW The abbreviation for kilowatt, which is a unit of measurement of electrical power. A kilowatt (KW) equals 1000 watts and is the product of the volts times the amperes divided by 1000 when used in rating direct current machinery. Also the term used to indicate true power in an ac circuit.

Kilowatt hour A kilowatt hour is the amount of electrical power represented by 1000 watts for a period of 1 hour. Thus a generator that delivers 1000 watts for a period of 1 hour would deliver 1 kilowatt hour of electricity.

Lally column Compression member consisting of a steel pipe filled with concrete under pressure.

Laminated wood Wood built up of plies or laminations that have been joined either with glue or with mechanical fasteners. Usually the plies are too thick to be classified as veneer and the grain of all plies is parallel.

Lighting outlet An outlet intended for the direct connection of a lampholder, lighting fixture, or pendant cord terminating in a lampholder.

Location, damp A location subject to a moderate amount of moisture, such as some basements, some barns, some cold-storage warehouses, and the like.

Location, dry A location not normally subject to dampness or wetness, a location classified as dry may be temporarily subject to dampness or wetness, as in the case of a building under construction.

Location, wet A location subject to saturation with water or other liquids, such as locations exposed to weather, washrooms in garages, and similar locations. Installations that are located underground or in concrete slabs, or masonry in direct contact with the earth shall be considered wet locations.

Logic The most reasonable arrangement of circuitry designed to accomplish certain objectives.

Low energy-power circuit A circuit that is not a remote-control or signal circuit but whose power supply is limited in accordance with the requirements of Class-2 remote-control circuits.

Modulation The process of varying the amplitude, frequency, or the phase of a carrier wave in accordance with specific data and information.

Multioutlet assembly A type of surface or flush raceway designed to hold conductors and attachment plug receptacles and assembled in the field or at the factory.

National electrical code The National Electrical Code is sponsored by the National Fire Protection Association and is the prime reference of all electrical workers for building construction. It is often referred to as the "NEC" or the "Code".

Nonautomatic Used to describe an action requiring personal intervention for its control.

OHM The ohm is the unit of measurement of electrical resistance and represents the amount of resistance that permits current to flow at the rate of one ampere under a pressure of one volt. The resistance (in ohms) equals the pressure (in volts) divided by the current (in amperes).

Outlet In the wiring system, a point at which current is taken to supply utilization equipment.

Outline lighting An arrangement of incandescent lamps or gaseous tubes to outline and call attention to certain features such as the shape of a building or the decoration of a window.

Oven, wall-mounted A domestic oven for cooking purposes designed for mounting into or onto a wall or other surface.

Panelboard A single panel or group of panel units designed for assembly in the form of a single panel; includes buses and may come with or without switches and/or automatic overcurrent protective devices for the control of light, heat, or power circuits of small individual as well as aggregate capacity. It is designed to be placed in a cabinet or cutout box placed in or against a wall or partition and accessible only from the front.

Pilaster Flat square column attached to a wall and projecting about a fifth of its width from the face of the wall.

Plate The principal anode in an electron tube to which the electron stream is attracted.

Plenum Chamber or space forming a part of an air-conditioning system.

Potential The difference in voltage between two points of a circuit. Frequently one is assumed to be ground (zero potential).

Potentiometer An instrument for measuring an unknown voltage or potential difference by balancing it, wholly or in part, by a known potential difference produced by the flow of known currents in a network of circuits of known electrical constants.

Power The rate of doing work or expending energy.

Power Factor When the current waves in an alternating current circuit coincide exactly in time with the voltage waves, the product of volts times amperes gives volt amperes which is true power in watts (or in KW if divided by 1000). When the current waves lag behind the voltage, due to inductive reactance (or lead due to capacitive reactance), they do not reach their respective values at the same time. Under such conditions, the product of volts and amperes does not give true average watts. Such a product is called volt amperes or apparent watts. The factor by which apparent watts must be multiplied to give the true watts is known as the power factor (PF).

Power factor depends on the amount of lag or lead, and is the percentage of apparent watts, which represents true watts. With a power factor of 80%, a fully loaded 5KVA alternator will produce 4 kW. When the rating of a power unit is stated in kVA at 80% PF, it means that with an 80% PF load, the generator will generate its rated voltage providing the load does not exceed the kVA rating. For example, the kVA rating of an engine-driven alternator with automatic voltage regulation is usually determined by the maximum current that can flow through the windings without injurious overheating or by the ability of the engine or other prime mover to maintain the normal operating speed. A resistance load such as electric lamp bulbs, irons, toasters, and similar devices is a unity power factor load. Motors, transformers, and various other devices cause a current wave lag which is expressed in the power factor of the load.

Precast concrete Concrete units (such as piles or vaults) cast away from the construction site and set in place.

Qualified person One familiar with the construction and operation of the apparatus and the hazards involved.

Raceway Any channel designed expressly for holding wire, cables, or bus bars and used solely for this purpose.

Raintight So constructed or protected that exposure to a beating rain will not result in the entrance of water.

Reactance Reactance is opposition to the change of current flow in an ac circuit. The rapid reversing of alternating current tends to induce voltages that oppose the flow of current in such a manner that the current waves do not coincide in time with the voltage waves. The opposition of self inductance to the flow of current is called "inductive reactance" and causes the current to lag behind the voltage producing it. The opposition of a condenser or of capacitance to the change of alternating current voltage causes the current wave to lead the voltage wave. This is called "capacitive reactance." The unit of measurement for either inductive reactance or capacitive reactance is the ohm.

Readily accessible Capable of being reached quickly, for operation or inspection, without requiring those to whom ready access is requisite to climb over or remove obstacles or resort to portable ladders, chairs, etc.

Receptacle (convenience outlet) A contact device installed at an outlet for the connection of an attachment plug.

Receptacle outlet An outlet in which one or more receptacles are installed.

Rectifiers Devices used to change alternating current to uni-directional current.

Relay An electromechanical switching device that can be used as a remote control.

Remote-control circuit Any electrical circuit that controls any other circuit through a relay or an equivalent device.

Resistance Electrical resistance is opposition to the flow of electric current and may be compared to the resistance of a pipe to the flow of water. All substances have some resistance but the amount varies with different substances and with the same substances under different conditions.

Resistor A resistor is a poor conductor used in a circuit to create resistance that limits the amount of current flow. It may be compared to a valve in a water system.

Resonance In a circuit containing both inductance and capacitance, a condition in which the inductive reactance is equal to and cancels out the capacitance reactance.

Riser Upright member of stair extending from tread to tread.

Roughing in Installation of all concealed electrical wiring; includes all electrical work done before finishing.

Saturation The condition existing in a circuit when an increase in the driving signal does not produce any further change in the resultant effect.

Sealed *(hermetic-type)* **motor compressor** A mechanical compressor consisting of a compressor and a motor, both of which are enclosed in the same sealed housing, with no external shaft or shaft seals, the motor operating in the refrigerant atmosphere.

Semiconductor A material midway between a conductor and an insulator.

Service The conductors and equipment used for delivering energy from the electricity supply system to the wiring system of the premises served.

Service cable The service conductors made up in the form of a cable.

Service conductors The supply conductors that extend from the street main or transformers to the service equipment of the premises being supplied.

Service drop The overhead service conductors from the last pole, or other aerial support, to and including the splices, if any, that connect to the service-entrance conductors at the building or other structure.

Service-entrance conductors, underground system The service conductors between the terminals of the service equipment and the point of connection to the service lateral.

Service equipment The necessary equipment, usually consisting of a circuit breaker, or switch and fuses, and their accessories, located near the point of entrance of supply conductors to a building and intended to constitute the main control and means of cutoff for the supply to that building.

Service lateral The underground service conductors between the street main, including any risers at a pole or other structure or from transformers, and the first point of connection to the service-entrance conductors in a terminal box, meter, or other enclosure with adequate space, inside or outside the building wall. Where there is no terminal box, meter, or other enclosure with adequate space the point of connection shall be considered to be the point of entrance of the service conductors into the building.

Service raceway The rigid metal conduit, electrical metallic tubing, or other raceway that encloses the service-entrance conductors.

Setting *(of circuit breaker)* The value of the current at which the circuit breaker is set to trip.

Sheathing First covering of boards or paneling nailed to the outside of the wood studs of a frame building.

Siding Finishing material nailed to the sheathing of a wood frame building and forming the exposed surface.

Signal circuit Any electrical circuit supplying energy to an appliance that gives a recognizable signal.

Single phase A single phase, alternating-current system has a single voltage in which voltage reversals occur at the same time and are of the same alternating polarity throughout the system.

Soffit Underside of a stair, arch, or cornice.

Solenoid An electromagnet having a movable iron core.

Soleplate Horizontal bottom member of wood-stud partition.

Studs Vertically set skeleton members of a partition or wall to which lath is nailed.

Switch, general-use A switch intended for use in general distribution and branch circuits. It is rated in amperes and is capable of interrupting its rated voltage.

Switch, general-use snap A form of general-use switch so constructed that it can be installed in flush device boxes or on outlet covers, or otherwise used in conjunction with wiring systems recognized by the Code.

Switch, ac general-use snap A form of general-use snap switch suitable only for use on alternating-current circuits and for controlling the following:

1. Resistive and inductive loads (including electric discharge lamps) not exceeding the ampere rating at the voltage involved.

2. Tungsten-filament lamp loads not exceeding the ampere rating at 120 volts.

3. Motor loads not exceeding 80% of the ampere rating of the switches at the rated voltage.

Switch, ac-dc general use snap A form of general use snap switch suitable for use on either direct- or alternating-current circuits and for controlling the following:

1. Resistive loads not exceeding the ampere rating at the voltage involved.

2. Inductive loads not exceeding one-half the ampere rating at the voltage involved, except that switches having a marked horsepower rating are suitable for controlling motors not exceeding the horsepower rating of the switch at the voltage involved.

3. Tungsten-filament lamp loads not exceeding the ampere rating at 125 volts, when marked with the letter T.

Switch, isolating A switch intended for isolating an electric circuit from the source of power. It has no interrupting rating and is in tended to be operated only after the circuit has been opened by some other means.

Switch, motor-circuit A switch, rated in horsepower, capable of interrupting the maximum operating overload current of a motor having the same horsepower rating as the switch at the rated voltage.

Switchboard A large single panel, frame, or assembly of panels, having switches, overcurrent and other protective devices, buses, and usually instruments, mounted on the face or back or both. Switchboards are generally accessible from the rear as well as from the front and are not intended to be installed in cabinets.

Synchronous Simultaneous in action and in time (in phase).

Tachometer An instrument for measuring revolutions per minute.

Thermal cutout An overcurrent protective device containing a heater element in addition to and affecting a renewable fusible member which opens the circuit. It is not designed to interrupt short-circuit currents.

Thermally protected *(as applied to motors)* Refers to the words "Thermally Protected" appearing on the nameplate of a motor or motor-compressor, and means that the motor is provided with a thermal protector.

Thermal protector *(as applied to motors)* A protective device assembled as an integral part of a motor or motor compressor and that, when properly applied, protects the motor against dangerous overheating due to overload and failure to start.

Three phase A three phase, alternating-current system has three individual circuits or phases. Each phase is timed so the current alternations of the first phase is $1/3$ cycle (120°) ahead of the second and $2/3$ cycle (240 °) ahead of the third.

Transformer A device used to transfer energy from one circuit to another. It is composed of two or more coils linked by magnetic lines of force.

Trusses Framed structural pieces consisting of triangles in a single plane for supporting loads over spans.

Utilization equipment Equipment that utilizes electric energy for mechanical, chemical, heating, lighting, or other similar useful purposes.

Ventilated Provided with a means to permit circulation of air sufficient to remove an excess of heat, fumes, or vapors.

Volt The practical unit of voltage or electromotive force. One volt sends a current of one ampere through a resistance of one ohm.

Voltage Voltage is the force—pressure or electromotive force (EMF)—that causes electric current to flow in an electric circuit. Its unit of measurement is the volt, which represents the amount of electrical pressure that causes current to flow at the rate of one ampere through a resistance of one ohm. Voltage in an electric circuit may be considered as being similar to water pressure in a pipe or water system.

Voltage drop The voltage drop in an electrical circuit is the difference between the voltage at the power source and the voltage at the point at which electricity is to be used. The voltage drop or loss is created by the resistance of the connecting conductors.

Voltage to ground In grounded circuits the voltage between the given conductor and that point or conductor of the circuit which is grounded; in ungrounded circuits, the greatest voltage between the given conductor and any other conductor of the circuit.

Watertight So constructed that moisture will not enter the enclosing case or housing.

Watt The watt is the unit of measurement of electrical power or rate of work. 746 watts is equivalent to 1 horse-power. The watt represents the rate at which power is expended when a pressure of one volt causes current to flow at the rate of one ampere. In a dc circuit or in an ac circuit at unity (100%) power factor, the number of watts equals the pressure (in volts) multiplied by the current (in amperes).

Weatherproof So constructed or protected that exposure to the weather will not interfere with successful operation.

Web Central portion of an I-beam.

Appendix
Electrical Specifications

Divisions 1 through 16 of the written specifications cover job requirements of a specific part of the construction work. Included in these divisions are the type and grade of materials to be used, equipment to be furnished, and the manner in which it is to be installed. Each division indicates the extent of the work covered and should be so written as to leave no doubt in anyone's mind about whether a certain part of the work is to be performed by a certain subcontractor and included in one section or another.

Division 16 of the specifications covers the electrical and related work on a given project, including the grade of materials to be used and the manner of installation for the electrical system. The following is an outline of the various sections normally included in Division 16—Electrical—of the written specifications for building construction.

16010 General Provisions The general provisions of the electrical specifications normally consist of a selected group of considerations and regulations that apply to all sections of the division. Items covered may include the scope of the work (work included and not included) in the electrical contract, electrical reference symbols, codes and fees, tests, demonstration of the completed electrical system, and identification of the equipment and components used in the installation.

16100 Basic Materials and Methods Definitive statements in this portion of the specifications should establish the means of identifying the type and quality of materials and equipment selected for use. This section should further establish the accepted methods of installing various materials such as raceways, conduits, bus ducts, underfloor ducts, cable trays, wires and cables, wire connections and devices, outlet boxes, floor boxes, cabinets, panelboards, switches and receptacles, motors, motor starters, disconnects, over-current protective devices, supporting devices, and electronic devices.

16200 Power Generation This section normally covers items of equipment used for emergency or standby power facilities, the type used to take over essential electrical service during a normal power source outage. This section usually cites requirements for a complete installation of all emergency circuits on a given project, including emergency service or standby power in the form of a generator set or storage batteries, automatic control facilities, feeders, panelboards, disconnects, branch circuits, and outlets.

Items to be fully described in this section include the generator and its engine (reciprocating or turbine), cooling equipment, exhaust equipment, starting equipment, and automatic or manual transfer equipment.

16300 Power Transmission Unlike the feeders, branch circuits, and the like that carry electrical power inside a building, high voltage (over 600 volts) power transmission is the subject of this section. Normally the specifications that require this section are for those projects constructed on government reservations and large industrial sites.

Cable and equipment specified in this section almost always is over 2.4 kV and includes such items as substations, switchgear, transformers, vaults, manholes, rectifiers, converters, and capacitors. Size and type of enclosures, as well as instrumentation, may also be included here.

16400 Service and Distribution Power distribution facilities (under 600 volts) for the project's service entrance, metering, distribution switchboards, branch circuit panelboards, feeder circuits, and the like are described in this section by paragraphs or clauses covering selected related equipment items.

In addition to the electric service characteristics—voltage, frequency, phase, etc., the quality and capacity levels of all items involved in the service entrance and power distribution system should be clearly defined. Typical items include the size and number of conductors, installation and supporting methods, location, rating, type and circuit protection features of all main circuit breakers, and other disconnecting means. The interrupting capacity of fuses is especially important and should be a major consideration in this section of the electrical specifications.

Other items for consideration are grounding, transformers (usually dry type), underground or overhead service, primary load interrupters, converters, and rectifiers.

16500 Lighting This section covers general conditions relating to selected lighting equipment to ensure that all such equipment is furnished and installed exactly as designated by the architect or engineer. Further clauses establish the quality and type of interior lighting fixtures, luminous ceilings, signal lighting, exterior lighting fixtures, stadium lighting, roadway lighting, accessories, lamps, ballasts and related accessories, poles, and standards. Methods of installation also are included in most sets of specifications.

Where special lighting equipment is specified, the specifications normally call for large, detailed shop drawings to be submitted to the architect or engineer for approval prior to installation.

16600 Special Systems Items that may be covered here include a wide variety of special systems unusual to conventional electrical installation. Examples of such items are lightning protection systems, special emergency light and power systems, storage batteries, battery charging equipment, and perhaps cathodic protection. However, this section is by no means limited to these few items. Many other special systems can be described in this section of the electrical specifications.

16700 Communications Equipment items that are interconnected to permit audio or visual contact between two or more stations or to monitor activity and operations at remote points are covered here. Most clauses deal with a particular manufacturer's equipment and state what items will be furnished and what is expected of the system once it is in operation.

Items covered under this section include radio, short-wave, and microwave transmission; alarm and detection systems; smoke detectors; clock and program equipment; telephone and telegraph equipment; intercommunication and public address equipment; television systems master TV antenna equipment; and learning laboratories.

16850 Heating and Cooling Because of working agreements among labor unions, most heating and cooling equipment is installed by workmen other than electricians, and the requirements are usually covered in Division 15—Mechanical—of the written specifications. In some cases, however, the electrical contractor is responsible for installing certain pieces of heating and cooling equipment, especially on residential and apartment projects.

The main point this section of the specifications should make is that the system installation meet with the design requirements. To do this, electric heat specifications should include pertinent data about the factors of the building insulation on which the design is based, as well as installation instructions for the selected equipment. To further aid in ensuring a proper installation, exact descriptions (manufacturer, catalog number, wattage rating, etc.) of the units are normally specified.

Items in this category of the electrical specifications include snow melting cables and mats, heating cable, electric heating coil, electric baseboard heaters, packaged room air conditioners, radiant heaters, duct heaters, and fan-type floor, ceiling, and wall heaters.

16900 Controls and Instrumentation As the name implies, this section covers all types of controls and instrumentation used on a given project. Examples include recording and indicating devices, motor control centers, lighting control equipment, electrical interlocking devices and applications, control of electric heating and cooling, limit switches, and numerous other such devices and systems.

Other divisions of the specifications besides Division 16, especially Division 15—Mechanical—may involve a certain amount of electrical work. The responsibility for such work should be clearly defined. Without great care by all concerned, confusion about responsibility may result in the electrical contractor paying for work he has assumed was not under his division (Division 16).

The most common sources of confusion are in control wiring for boilers, heating, ventilating, and air conditioning systems; control, signal, and power wiring beyond the machine room; disconnect switches for elevator construction; automatic machinery controls; wiring on machine tools; wiring on overhead cranes and hoists; mounting and connecting of motors; connecting hospital, laundry, and restaurant equipment; connecting electric signs; connecting motion picture projection and sound equipment; installing electric lighting fixtures furnished by someone other than the electrical contractor; connecting unit heaters, unit ventilators, electric fans, electric water heaters, electric water coolers, electric ranges, and other appliances when they are not furnished by the electrical contractor; connecting transformers, and similar items.

The following electrical specification is typical of those used on small to medium size electrical construction work. It is presented to show the electrical inspector how such specifications are worded.

DIVISION 16
SECTION 16B
ELECTRICAL DETAIL SPECIFICATIONS

16B.01 *GENERAL*

See Section 15A, GENERAL SPECIFICATIONS FOR MECHANICAL AND ELECTRICAL WORK, of these Specifications for General Specifications relating to ELECTRICAL WORK.

16B.02 *SCOPE OF WORK*

Every item of labor, material, devices, and appurtenances for installing a complete electrical system and other related systems as specified herein and shown on the drawings.

16B.03 *WORK NOT INCLUDED*

Certain electrical equipment will be provided in place by another Contractor ready for connection by the Electrical Contractor. Other pieces of equipment will be provided f.o.b. premises, which the Electrical Contractor shall mount and connect to electrically.

16B.04 *REGULATIONS AND STANDARDS*

(a) The completed installation and all materials and equipment shall conform to all applicable portions of the National Fire Protection Association Codes and all other governing codes and regulations.

(b) The above standards are intended as a minimum and shall be exceeded if required by the specifications, the drawings, or governing regulations. In the event of conflict between the above mentioned codes and regulations and information contained in the contract documents, the applicable code or regulation shall take precedence and no extra compensation will be allowed due to such conflict.

16B.05 THIS NUMBER NOT USED.

16B.06 *SHOP DRAWINGS*

Submittal data is required for items in this section

marked by asterisks and as explained by the GENERAL SPECIFICATIONS. Submittal data is required for all proposed substitutions.

16B.07 *INSTALLATION OF WORK*

(a) *Pull Boxes:* Provide all necessary pull boxes and cable raceways where indicated or required. Boxes and raceways to be constructed of code gauge galvanized steel with adequate braces and supports. Certain pull boxes may be shown on the drawings for specific design reasons but this is not to preclude the fact that additional boxes will be required to conform with codes and good practice. Where flush mounted boxes are required, they shall have a 3/4″ overlapping trim and countersunk for flat head screws. Other boxes may have round head screw or hinged covers.

(b) *Supports:* Provide metallic supports as required for the proper installation of raceway systems and all other equipment installed under this contract.

(c) *Empty Raceways:* Pull #12 gauge galvanized fishing wires through all empty conduits for all systems. These wires to remain in the conduits.

16B.08 *TESTING*

(a) *System Readings:* The Contractor shall take certain voltage and current readings, record all values and submit in triplicate to the Engineer. Two complete sets of readings are required, one under no load and one under maximum available load. The current and voltage shall be recorded on each phase (plus voltage between all phases) at main panelboard and at each branch circuit panelboard. Additional spot readings shall be made if required. Resistance of grounding conductors shall be tested and recorded. Forms for submitting this report may be obtained from the Engineer's office. A sample form is bound herewith.

(b) *Equipment Readings:* The Electrical Contractor shall take voltage and amperage readings on each phase of each motor circuit and each resistance heater circuit installed under this Contract and record same. Also record motor nameplate data, actual motor heater protective device ratings, and all other data necessary for selection of heater device.

16B.09 *WIRING METHODS*

(a) *General:* All wires for all systems shall be installed in rigid metal raceways and terminated in boxes or cabinets unless otherwise specified herein.

(b) *Electrical Metallic Tubing:* E.M.T. in sizes 4″ and smaller will be acceptable for all locations, except underground, in exposed locations (except exposed on ceilings or suspended reasonably close to ceilings) and where cast in concrete.

(c) *Flexible Metal Conduit:* Use for short motor connections and where permitted by the Engineer.

(d) *Installation:* Provide conduit systems to achieve required distribution and control. Conduit runs are not shown on the drawings unless specifically noted or indicated. No rigid raceway for line voltage wiring shall be smaller than 3/4″ (except for flexible conduit) unless specifically indicated otherwise. The Contractor shall size all other raceways based on the N.E.C. and verify the sizes shown on the drawings, increasing same if smaller than permissible by code.

(e) *Materials:*

1. Rigid Metal Conduits: Of mild steel tube, electro or hot-dipped galvanized, as manufactured by Pittsburgh Standard, Youngstown, Republic, or Steelduct. Couplings shall be hot-dipped galvanized or sheradized steel. Exposed threads not otherwise protected shall be painted after installation with two coats of asphaltum paint if concealed and two coats of lead and oil paint if installed in an exposed location.

2. Electrical Metallic Tubing: Of mild steel tube in sizes 3/4″ to 4″, hot-dipped galvanized or electro-galvanized and U.L. labeled. Republic, Steelduct or accepted equal.

3. Flexible Metal Conduit: Liquidtight, Anaconda "Sealtite", or accepted equal (grounded) for all sizes

through 3″. Use for all motor connections and all other equipment subject to movement or vibration. Galvanized single strip steel, if accepted by the Engineer, may be used for other locations and for sizes above 3″.

(f) *Fittings:*

1. For rigid conduit: Couplings to be galvanized or sheradized steel. Double galvanized steel locknuts shall be used where required by code. Single locknut and bushing may be used elsewhere. O.Z. type B (or BLG where required), or accepted equal insulating bushings to be installed on all rigid conduit terminations where bushings are required or would normally be used.

2. For E.M.T.: T & B, or accepted equal, steel set screw connectors, permanently indented or gland compression type.

3. For Flexible Metal Conduit: Standard liquidtight fittings by T & B or accepted equal for liquidtight conduit. Standard steel flexible metal conduit fittings by T & B or accepted equal for standard flexible metal conduit.

4. Conduit "L's": Galvanized steel, threaded, Crouse-Hinds, or accepted equal, "LB" or "LBD".

5. Grounding Clamps: T & B #3902 series or #3903 series, or accepted equal, as required by water pipe size and/or grounding conductor size. Main service ground clamp shall be installed on the street side of the main cold water service valve

6. Cable Supports: O.Z. type "M", or accepted equal. To be installed for the support of cables, #6 and larger at the termination of any vertical run more than 6′-0″.

7. Underground Raceway Seals: Provided on each underground electrical raceway entering building. Mechanical type by O.Z. or accepted equal.

(g) *Installation:* Conceal all raceways except where specifically noted on drawings or permitted as exposed. Runs in Mechanical Room Areas may be exposed.

1. Rigid Metal Conduit: All joints shall be properly threaded and made tight in standard conduit couplings. All thread conduits or nipples are not acceptable. All elbows and nipples to be same material as specified for conduit. Cap raceway systems during course of construction and clean inside before installation of conductors. Exposed conduit must be run parallel with the building walls and supported in a neat, substantial manner. Clamps to be malleable galvanized iron and hangers to be rod type steel. Wire (galvanized 12 gauge) may be used only for tying to steel joists and then only in concealed spaces.

2. Electrical Metallic Tubing: Applicable specifications for installation of rigid metal conduit applies to E.M.T.

(h) *Exposed Masonry:* Where wall finish is exposed masonry, outlets and raceways shall be so placed in wall that the masonry unit can be neatly set around it with minimum cutting and without injury to the exposed masonry face. Dimensioned heights of outlets in exposed masonry are intended to mean to the closest masonry joint.

(i) *Coordination:* This Contractor shall refer to all other drawings to prevent conflicts between electrical and other work. Ductwork and piping shall take precedence over conduit runs which shall be installed so that conflicts do not occur.

(j) *Outlet Boxes:* Boxes must be set flush or set to meet N.E.C. requirements; otherwise box extensions shall be installed.

16B.10 *CONDUCTORS*

(a) *General:* All wires and cables shall be as manufactured by Anaconda, Phelps Dodge, Okonite, Simplex, Rome, or accepted equal.

(b) *Conductor Ties:* Inside each enclosure other than outlet and junction boxes, conductors shall be bundled and trained utilizing T & B "Ty-Rap", or accepted equal ties.

(c) *Conductor Sizes:*

1. Line Voltage Wiring: No. 12 AWG minimum unless

indicated otherwise. Larger sizes as required or shown on drawings.

2. Control Wiring: If not carrying motor current, No. 14 AWG unless otherwise indicated or required by load encountered.

(d) *Stranding:* All conductors No. 8 and larger.

(e) *Insulation:* THW unless otherwise indicated on drawings.

(f) *Conductor Material:* All shall be copper. The Contractor shall have the option to use aluminum conductors as specified hereinafter.

16B.11 *CONNECTORS AND LUGS*

(a) *Material:* Copper or suitable copper alloy for all current carrying parts and all parts coming in contact with conductors.

(b) *Connectors and Lugs, No. 8 and Larger Conductors:* Compression type T & B "Color-Keyed" or accepted equal.

(c) *Connectors, No. 10 and Smaller Conductors:* Permanently indented pressure connectors T & B, Buchanan or accepted equal. "Scotchloks" by the 3M Company or "Wing-Nut" by Ideal, acceptable.

(d) *Lugs No. 10 and Smaller Conductors:* Permanently indented or compression type by Buchanan, Burndy or T & B. Washer head screw terminals without lugs are acceptable on neutral bars, circuit breakers, wiring devices and other equipment unless otherwise indicated.

16B.12 *OUTLET BOXES*

(a) *General:* Sheradized or galvanized (after fabrication) sheet steel boxes, Appleton, RACO, Steel City or accepted equal. All boxes shall have ears turned in. Multiple gang boxes must be one piece type (not built-up). Provide 3/8" (or larger if required) fixture stud in all fixture boxes. Provide appropriate covers as required, including 3/4" deep plaster ring covers where plaster may be encountered. Provide vaporproof outlet boxes for vaporproof fixtures. Size and type of boxes as required by location and N.E.C., except where exposed masonry occurs, use one piece "tile boxes". Use 4" square boxes with "tile rings" where larger sizes are required. All outlet boxes requiring hangers shall be hung with metal hangers by Appleton Electric Mfg. Co., or accepted equal. All exposed boxes below ceiling level shall be cast type FS or FD.

(b) *Heights of Outlets:* Shown on the drawings, specified herein, or as permitted on the job.

(c) *Floor Boxes:* Unless noted otherwise, Steel City #601AL-P-60-DU-AL, or accepted equal, (or terrazzo type where needed) cast iron watertight floor box, aluminum cover and flange, with outlets as specified under "WIRING DEVICES". Each box shall have two #703-AL nozzles. Other floor boxes as specified elsewhere or called for on the drawings. Provide carpet flanges where carpet is encountered.

16B.13 *WIRING DEVICES*

(a) *General:* Switches, receptacles, and other devices to be as indicated on drawings and as manufactured by Hubbell, Bryant, P & S, or accepted equal.

(b) *Pilot Lights:* Neon type with amber natural color glass, unless otherwise called for. Lamp to be sized to adequately illuminate the jewel.

(c) *Plates:* For all systems unless indicated otherwise.

1. For Concealed Work: Type marked on drawings.

2. For Exposed Work: Cast FS or FD covers with gaskets.

3. Screws: Heads to match plate material and finish.

16B.14 *DISCONNECT SWITCHES, FUSES, MOTOR CONTROLLERS*

(a) *Disconnect Switches:* Rated for voltage encountered, poles and amperage as required. Normal Duty, NEMA Type 1 enclosures, fusible unless indicated, Square-D, G.E., Westinghouse, or accepted equal. All to be products of the same manufacturer. Provide raintight or explosion-proof switches where required.

(b) *Fuses:* Dual element, time-lag, cartridge type, Fusetrons by Bussman, or accepted equal. Fuses for motor circuits shall be sized in accordance with Bussman sizing chart for motors unless otherwise required for a specific motor. Labels indicating size and type of replacement fuses shall be glued to inside of door on all fusible switches and panels.

(c) *Motor Controllers:* Furnished by the Electrical Contractor unless indicated otherwise.

1. Motor Protective Switches: Marked M.P. Sw. on drawings. Manual type with overload relay for each phase and with poles as required.

2. Magnetic Starters: Marked M.S. on drawings. Across-the-line, magnetic type unless indicated otherwise. Overload relay for each phase and with poles as required. Where Comb. M.S. is called for, provide fusible disconnect in starter unless nonfusible is noted. All to have 120V. control circuits and holding coils. Provide individual control transformers of required size where higher voltage is encountered.

3. Heaters: Install thermal overload heater elements in all switches and starters on the job whether or not the switches and starters are furnished by this subcontractor. Assume responsibility for proper application of motor running protection for all motors in accordance with the manufacturer's recommendations and the nameplate rating of the motors actually installed. All phased to have overload protection.

4. Pushbutton Stations: Standard duty pushbutton stations with maintained contact, START-STOP buttons and amber indicating light unless otherwise indicated.

5. Enclosures: The enclosure shall be surface mounted NEMA-1 unless otherwise indicated. Hand-off-automatic switches and pilots in covers as required.

6. Manufacturers: Furnas, Square-D, Clark Controller, Westinghouse, Allen-Bradley or accepted equal. All controllers and accessories to be by the same manufacturer, ________, or approved equal.

16B.15 *MECHANICAL EQUIPMENT WIRING*

(a) General: This Contractor shall consult all other sections of these specifications containing any type of equipment requiring electrical connections and allow for wiring and controlling all equipment as described therein even though not shown on the electrical drawings.

(b) *Motors:* All motors and motordriven equipment will be furnished and installed in place by another subcontractor, ready for connection by this Contractor with control equipment provided by the Contractor who furnishes the equipment. Starters and switches with all accessories to be furnished by this Contractor unless indicated otherwise. Motor branch circuits, controllers, switches, etc., are designed for motor sizes as shown on the drawings. The motors and motordriven equipment actually installed may vary from these sizes. This Contractor shall check sizes and notify the Engineer of any required changes.

(c) *Responsibility:* The Mechanical Contractor is responsible for the proper operation of the mechanical systems, but this Contractor is responsible for the electrical work in connection therewith.

16B.16 *PANELBOARDS***

(a) *General:* Dead front automatic circuit breaker or fusible type as indicated and as manufactured by Kinney Electrical Mfg. Co., Westinghouse, G.E., or accepted equal, and designed for type of service encountered.

(b) *Breakers:* Westinghouse, or accepted equal, whose frame sizes are called for on the drawings. The breaker's line terminal connection shall bolt directly to the panelboard's bus connecting strap. Mult-pole breakers shall be in one molded case, common trip, with one operating handle. Where space only is indicated, provide mounting straps, for future breaker.

(c) *Miscellaneous:* Cabinets of all welded code gauge galvanized steel, 6″ minimum outside depth, with studs for mounting interiors. Minimum of 4″ side gutters; top and bottom gutters as required; special gutter sizes as indicated on the drawings. Raceway openings to be 2″ clear from front of cabinet. Provide solid neutral bus. Steel front to be grey lacquered code gauge having index card holder, flush latch and lock (two keys for each panel) and adjustable trim clamps. Index cards to be properly labeled with typewritten letters. All bus to be copper with silver, cadmium or tin plating. Fusible panels to be fused with Fusetrons.

16B.17 *LIGHTING FIXTURES***

(a) *General:* Furnish and install, complete with lamps, all lighting fixtures indicated on drawings. The lamp wattage is indicated as a prefix to the fixture type number.

(b) *Basis for Bid:* Manufacturers' names and catalog numbers are used to designate the type, general design, size, quality, etc., of fixtures desired. Fixtures by other manufacturers will be considered for acceptance after award of contract.

(c) *Fluorescent:*

1. Ballasts: Jefferson, G.E., or Universal having 90 per cent or higher power factor and bearing C.B.M., E.T.L., and U/L Labels. Sound rating equal to G.E. rating A and all to be class "P".

2. Lamps: G.E. or Westinghouse, Standard Warm White, or accepted equal.

(d) *Incandescent Lamps:* G.E. or Westinghouse general lighting line, I.F., *130 volt* unless otherwise specified. Quartz lamps for voltage encountered.

(e) *Hanging:* All fixtures to be adequately supported by fixture studs, conduit stems, steel rods, or bar hangers.

(f) *Plaster Frames:* Recessed fixtures in plaster ceiling areas shall be installed in rolled metal plaster frames furnished and installed by the Electrical Contractor.

(g) *Ceiling Types:* Before fixture order is placed, the Contractor will be responsible for checking type of ceiling in all spaces to determine if fixture type called for is correct to fit ceiling encountered.

16B.18 *WIRING FOR MOTION PICTURE SOUND EQUIPMENT*

(a) *Wires and Conduits:* Extend 3/4" conduit and two (2) #12 wires between the outlets shown on drawings.

(b) *Receptacles:* Switchcraft #11/280 female receptacle mounted through a hole in a blank plate for wall location and mounted in box similar to that specified for telephones for floor location.

(c) *Grounding:* Positively ground to the power and lighting conduit system.

16B.19 *PHOTO-CELLS AND RELAYS*

Provide photo-cells, time switches, relays, and contactors as shown on drawings.

16B.20 *SERVICE*

(a) The Power Company will furnish and install the underground primary service wire and the pad mounted transformer. This Contractor shall furnish and install the secondary service.

(b) Secondary Service will be *120/208y* volts, *three* phase, *four* wire, 60 Hertz A. C.

(c) *Metering* will be by the Power Company. Provide one 1–1/4" empty conduit from C.T. Cabinet to the meter cabinet. The Power Company will furnish the meter cabinet and C.T. Cabinet which will be mounted by the Electrical Contractor. The Power Company will run control wires to the meter. The C.T.'s will be furnished by the Power Company and will be installed by the Electrical Contractor.

(d) *Grounding:* Ground service in accordance with the N.E.C. and as indicated in the power-rise diagram.

16B.21 *ALUMINUM CONDUCTOR OPTION*

(a) Drawing E-2 shall be modified to include the following Contractor's option:

The Contractor shall have the option of using 600 V, Type THW, aluminum wires for all feeders, and branch circuit wires #8 AWG and larger sizes. All shall be in accordance with the revised schedule of wire and conduit quantities and sizes which follows. If this option is excercised, all panelboard and breaker lugs shall be changed to conform with the revised quantities and sizes of conductors indicated and of an approved type to accept aluminum conductors and attach to copper bus. All panel gutters shall be increased in size as deemed advisable to receive the additional conductors and lugs. Pull boxes and junction boxes shall be increased as necessary to receive the additional conduits and wires. The Contractor shall state on the Bid Form as to whether or not this option will be exercised.

(b) Schedule of wire and conduit for feeders and branch Circuits #6 AWG and larger if 600 volt type THW aluminum conductors are used

Feeder or Branch Crt.	No. Wires	Size Wire	Size Cond.
MAIN SERVICE	8	#500 MCM	2 @ 3-½″
MP-2	3	#4	1-¼″
	1	#12	
MP-3	3	#1/0	2″
	1	#12	
MP-5	4	#250 MCM	2-½″
MP-6	4	#250 MCM	2-½″
MP-7	4	#250 MCM	2-½″
B-37	4	#4	1-¼″
C-19	3	#4	1-¼″
	1	#12	

Index